The (Fabulous) FIBONACCI Numbers

"I couldn't imagine a more lucid and entertaining introduction to these fabulous Fibonacci numbers. This book provides evidence of the beauty of mathematics through this amazing phenomenon that seems to permeate just about everything—both in- and outside of the world of mathematics. Read the book and you will find a new opportunity and a surprise for some folks, to love the world of the mathematicians!"

Charlotte K. Frank, PhD
Senior Vice President—Research and Development
McGraw-Hill Education

"A book about mathematics which intrigues and delights, which draws the reader into a mystery as good as any detective novel, is indeed to be welcomed. This book will appeal not only to the adult mathematician but also to young people who may never have thought of math as related to some of the most fundamental laws of nature. In clear and simple language, tables, and diagrams, Posamentier and Lehmann have successfully shown how the famous Fibonacci numbers offer both an explanation of our world, and ways of applying the numbers to everyday matters and objects. Given the crisis we face in math teaching, where a dangerously small number of young people are turning to math as a specialty, this book offers a real way of making mathematics a source of pleasure and a challenge. It provides a route to draw more people of every age into the magic of the numbers which shape the world we inhabit."

Baroness Perry of Southwark
Former president of Lucy Cavendish College
in the University of Cambridge,
formerly Her Majesty's Chief Inspector in the UK,
and former president of South Bank University, London

"This book is an excellent example of the elegance of mathematics, or in the words of Keats: 'truth is beauty.'"

Dr. Zeev Dagan
Provost, City College of New York

"This book takes the reader on a wonderous adventure into the Fibonacci numbers. It traces the rich history of these numbers and their surprising appearance in the world around us. Both the math novice and the math enthusiast will be enthralled as the authors reveal the secrets surrounding the fabulous Fibonacci numbers!"

Daniel Jaye
Assistant principal (retired), Stuyvesant High School
and director of academy programs, Bergen County Academies

"Zero, and one, and simple addition: this is the starting point of a most adventurous journey through a fascinating math-magic land. A. Posamentier and I. Lehman prove to be experienced tour guides, always showing unexpected relations to art, music, history."

Rudolf Taschner
Founder of the Math Space Museum in Vienna

The (Fabulous) FIBONACCI Numbers

Alfred S. Posamentier
Ingmar Lehmann

Afterword by
Herbert A.
Hauptman
Nobel Laureate

Prometheus Books
59 John Glenn Drive
Amherst, New York 14228–2119

Published 2007 by Prometheus Books

Inquiries should be addressed to
Prometheus Books
59 John Glenn Drive
Amherst, New York 14228–2119
VOICE: 716–691–0133, ext. 210
FAX: 716–691–0137
WWW.PROMETHEUSBOOKS.COM

15 14 13 12   6 5

Library of Congress Cataloging-in-Publication Data

Posamentier, Alfred S.
    The fabulous Fibonacci numbers / by Alfred S. Posamentier and Ingmar Lehmann.
        p. cm.
    Includes bibliographical references and index.
    ISBN 978–1–59102–475–0 (alk. paper)
    1. Fibonacci numbers. I. Lehmann, Ingmar. II. Title.

QA241.P665 2007
512.7'2—dc22

                                        2006035406

Printed in the United States of America on acid-free paper

To Barbara for her support, patience, and inspiration.

To my children and grandchildren: David, Lisa, Danny, Max, and Sam, whose futures are unbounded.

And in memory of my beloved parents, Alice and Ernest, who never lost faith in me.

Alfred S. Posamentier

To my wife and life partner, Sabine, without whose support and patience my work on this book would not have been possible.

And to my children and grandchildren: Maren, Claudia, Simon, and Miriam.

Ingmar Lehmann

Contents

Acknowledgments   **9**

Introduction   **11**

Chapter 1:   A History and Introduction to the
                 Fibonacci Numbers   **17**

Chapter 2:   The Fibonacci Numbers in Nature   **59**

Chapter 3:   The Fibonacci Numbers and the
                 Pascal Triangle   **77**

Chapter 4:   The Fibonacci Numbers and the
                 Golden Ratio   **107**

Chapter 5:   The Fibonacci Numbers and
                 Continued Fractions   **161**

Chapter 6:   A Potpourri of Fibonacci Number
                 Applications   **177**

Chapter 7:   The Fibonacci Numbers Found in Art
                 and Architecture   **231**

Chapter 8:   The Fibonacci Numbers and
                 Musical Form   **271**

Chapter 9:   The Famous Binet Formula for Finding
a Particular Fibonacci Number   **293**

Chapter 10:   The Fibonacci Numbers and
Fractals   **307**

Epilogue   **327**

Afterword by Herbert A. Hauptman   **329**

Appendix A: List of the First 500 Fibonacci Numbers,
with the First 200 Fibonacci Numbers
Factored   **343**

Appendix B: Proofs of Fibonacci Relationships   **349**

References   **371**

Index   **375**

Acknowledgments

The authors wish to thank Professor Stephen Jablonsky of the City College of New York–CUNY, who used his multifaceted talents to present the unusual appearances of the Fibonacci numbers in the realm of music from the design of compositions to the design of musical instruments.

Professor Ana Lucía B. Dias deserves our sincere thanks for her broad discussion of the occurrence of the Fibonacci numbers in the field of fractals.

Our dear friend Dr. Herbert A. Hauptman, the acknowledged first mathematician to win the Nobel Prize (Chemistry, 1985), provided a stimulating afterword for the book. This capstone piece will provide a challenge to the reader and may even have some elements never before discovered.

It was during a conversation with our friend James S. Tisch, president and CEO of the Loews Corporation, that he mentioned the applications of the Fibonacci numbers in the world of economics. He provided much information, and we took it the rest of the way to the pages of this book. For this, we thank him.

Dr. Lehmann acknowledges the occasional support of Tristan Vincent in helping him identify the right English words to best express his ideas.

We wish to also thank Professor Andreas Filler, of the Pädagogische Hochschule at Heidelberg (Germany); Heino Hellwig, Humboldt University, Berlin; and Hans-Peter Lüdtke, at the Heinrich-Hertz Gymnasium, Berlin, for providing some valuable comments during the writing of this book.

The success of a book is usually dependent on two types of support: our faithful guide through the development of the project for which we thank Linda Greenspan Regan, who gave us constant advice to make the book as appealing to the general readership as possible, and the content editor, Peggy Deemer, who, as usual, did a magnificent job editing the manuscript—no mean feat, given the complexity of the subject.

Naturally, we wish to thank, respectively, Barbara and Sabine, for their encouragement, patience, and support during our work on this book.

Introduction

The Fabulous Fibonacci Numbers

In a remote section of the Austrian Alps, there is a long-abandoned salt mine entrance with a cornerstone bearing the inscription "anno 1180." It refers to the year that the mine was established. Clearly there is something wrong with this designation. Scholars have determined that the first published use of the Hindu numerals (our common numerals) in the Western world was in 1202. It was in this year that Leonardo of Pisa (Leonardo Pisano), more commonly known as Fibonacci (pronounced: fee-boh-NACH-ee), published his seminal work *Liber Abaci*, or "book of calculation." He began chapter 1 of his book with:

> The nine Indian[1] figures are: 9 8 7 6 5 4 3 2 1.
> With these nine figures, and with the sign 0, which
> the Arabs call zephyr, any number whatsoever is written.

This constitutes the first formal mention of our base-ten numeral system in the Western world. There is speculation, however, that Arabs already introduced these numerals informally and locally in Spain in the second half of the tenth century.

Unlike luminaries from the past, whose fame today is largely based on a single work, such as Georges Bizet (1838–1875) for *Carmen*, Engelbert Humperdinck (1854–1921) for *Hänsel und*

---

1. Fibonacci used the term Indian figures for the Hindu-Arabic numerals.

11

Gretel, or J. D. Salinger (b. 1919) for *The Catcher in the Rye*, we should not think of Fibonacci as a mathematician who is known only for this now-famous sequence of numbers that bears his name. He was one of the greatest mathematical influences in Western culture and unquestionably the leading mathematical mind of his times. Yet it was this sequence of numbers, emanating from a problem on the regeneration of rabbits, that made him famous in today's world.

Fibonacci was a serious mathematician, who first learned mathematics in his youth in Bugia, a town on the Barbary Coast of Africa, which had been established by merchants from Pisa. He traveled on business throughout the Middle East and along the way met mathematicians with whom he entered into serious discussions. He was familiar with the methods of Euclid (ca. 300 BCE) and used those skills to bring to the European people mathematics in a very usable form. His contributions include introducing a practical numeration system, calculating algorithms and algebraic methods, and a new facility with fractions, among others. The result was that the schools in Tuscany soon began to teach Fibonacci's form of calculation. They abandoned the use of the abacus, which involved counting beads on a string and then recording their results with Roman numerals. This catapulted the discipline of mathematics forward, since algorithms were not possible with these cumbersome symbols. Through his revolutionary book and other subsequent publications, he made dramatic changes in western Europe's use and understanding of mathematics.

Unfortunately, Fibonacci's popularity today does not encompass these most important innovations. Among the mathematical problems Fibonacci poses in chapter 12 of *Liber Abaci*, there is one about the regeneration of rabbits. Although its statement is a bit cumbersome, its results have paved the way for a plethora of monumental ideas, which has resulted in his fame today. The problem, which is stated on page 25 (figure 1-2), shows the monthly count of rabbits as the following sequence of numbers: 1, 1, 2, 3, 5, 8, 13, 21, 34, 55, 89, 144, 233, 377, . . . , which is known today as the Fibonacci numbers. At first sight one may wonder what makes this sequence of numbers so revered. A quick inspection shows that this sequence of numbers can go on infinitely, as it begins with two 1s and continues to get succeeding terms by adding, each time, the last two numbers to get the next number (i.e., $1 + 1 = 2$, $1 + 2 = 3$, $2 + 3 = 5$, and so on). By itself, this is not very impressive. Yet, as you will see, there are no numbers in all of mathematics as ubiquitous as the Fibonacci numbers. They appear in geometry, algebra, number theory, and many other branches of mathematics. However, even more spectacularly, they appear in nature; for example, the number of spirals of bracts on a pinecone is always a Fibonacci number, and, similarly, the number of spirals of bracts on a pineapple is also a Fibonacci number. The appearances in nature seem boundless. The Fibonacci numbers can be found in connection with the arrangement of branches on various species of trees, as well as in the number of ancestors at every generation of the male bee on its family tree. There is practically no end to where these numbers appear.

Throughout these pages we will explore many manifestations, so that you may be motivated to find other instances in nature where these Fibonacci numbers surface. The book will also provide you with a gentle, yet illuminating discussion of the unusual nature of these numbers. Their interrelationships with other seemingly completely unrelated aspects of mathematics will open the door to applications in a variety of other fields and areas as remote as the stock market.

It is our desire that this book serve as your introduction to these fabulous numbers. We will provide you with the development of the Fibonacci numbers, a sort of history, one might say. Then we will investigate the many "sightings" of these numbers by subject area. For example, in geometry we will explore their relationship with the most beautiful ratio, known as the "golden ratio."[2] Here the Fibonacci numbers, when taken as quotients in consecutive pairs, approach the golden ratio:

$$\phi = 1.61803398874989484482045868343656\ldots$$

---

2. The golden ratio will be discussed in reasonable depth (with ample applications in geometry and art) so that the reader will have a true appreciation for it and, consequently, its relationship to the Fibonacci numbers. It is often symbolized by the Greek letter ϕ.

The larger the Fibonacci numbers, the closer their quotient approaches the golden ratio.

Consider, for example, the quotient of the relatively small pair of consecutive Fibonacci numbers:

$$\frac{13}{8} = 1.625$$

Then consider the quotient of the somewhat larger pair of consecutive Fibonacci numbers:

$$\frac{55}{34} = 1.61\overline{7647058823529411}7 \quad [3]$$

and the quotient of an even larger pair of consecutive Fibonacci numbers:

$$\frac{144}{89} = 1.61\overline{797752808988764044943820224719101123595505}$$

Notice how these increasingly larger quotients seem to surround the actual value of the golden ratio, ϕ. Then, when we take much larger pairs of consecutive Fibonacci numbers, their quotients get us ever closer to the actual value of the golden ratio, ϕ. For example,

$$\frac{4{,}181}{2{,}584} = 1.6180340557275541795665634674923\ldots$$

$$\frac{165{,}580{,}141}{102{,}334{,}155} = 1.6180339887498948909091006809994180339 88\ldots \quad [4]$$

---

3. The bar above the last repeat of 16 indicates that the 16 digits repeat indefinitely.
4. The fortieth Fibonacci number is 102,334,155 and the forty-first is 165,580,141.

Compare this last quotient to the value of the golden ratio:

$$\phi = 1.6180339887498948482045868343656\ldots$$

We will also discuss the nature and glory of the golden ratio itself. Its appearance in architecture and art is more than coincidental. If you were to sketch a rectangle encasing the front view of the Parthenon in Athens, Greece, you will have drawn a golden rectangle—that is, one whose length and width are in the golden ratio. Many artists have employed the golden rectangle in their art. The painting of *Adam and Eve* by the famous medieval German artist Albrecht Dürer (1471–1528), for instance, has the subjects encased in a golden rectangle.

Interestingly enough, the Fibonacci numbers did not get their prominence, and their name, until Edouard Lucas (1842–1891), a French mathematician, studied them in the second half of the nineteenth century. He mused about the start of the sequence, wondering what would happen if the sequence had begun with 1 and 3, rather than 1 and 1. So he studied this new sequence (following the same additive rule) and then compared it to the Fibonacci sequence. The Lucas numbers, which are 1, 3, 4, 7, 11, 18, 29, 47, 76, 123, . . . , interrelate with the Fibonacci numbers. We will explore their relationship later.

There is almost no end to the places where the Fibonacci numbers appear and their many applications. We will present mathematical recreations, as well as more serious properties of these numbers—all of which will fascinate the uninitiated as well as the math-savvy reader. You will be truly awed by these magnificent numbers and most likely will be continuously consciously seeking your own sightings of the Fibonacci numbers. Throughout this book we hope to appeal to all types of readers, but always have the general reader in mind. For the more mathematically advanced reader, however, we provide an appendix with proofs of statements made throughout this book. Our goal is to evoke the power and beauty of mathematics for all readers.

Chapter 1

A History and Introduction to the Fibonacci Numbers

With the dawn of the thirteenth century, Europe began to wake from the long sleep of the Middle Ages and perceive faint glimmers of the coming Renaissance. The mists rose slowly as the forces of change impelled scholars, crusaders, artists, and merchants to take their tentative steps into the future. Nowhere were these stirrings more evident than in the great trading and mercantile cities of Italy. By the end of the century, Marco Polo (1254–1324) had journeyed the Great Silk Road to reach China, Giotto di Bondone (1266–1337) had changed the course of painting and freed it from Byzantine conventions, and the mathematician Leonardo Pisano, best known as Fibonacci, changed forever Western methods of calculation, which facilitated the exchange of currency and trade. He further presented mathematicians to this day with unsolved challenges. The Fibonacci Association, started in 1963, is a tribute to the enduring contributions of the master.

Leonardo Pisano—or Leonardo of Pisa, Fibonacci[1]—his name as recorded in history, is derived from the Latin "filius Bonacci," or a son of Bonacci, but it may more likely derive from "de filiis

---

1. It is unclear who first used the name Fibonacci; however, it seems to be attributed to Giovanni Gabriello Grimaldi (1757–1837) at around 1790, or to Pietro Cossali (1748–1815).

Bonacci," or family of Bonacci. He was born to Guilielmo (William) Bonacci and his wife in the port city of Pisa, Italy, around 1175, shortly after the start of construction of the famous bell tower, the Leaning Tower of Pisa. These were turbulent times in Europe. The Crusades were in full swing and the Holy Roman Empire was in conflict with the papacy. The cities of Pisa, Genoa, Venice, and Amalfi, although frequently at war with each other, were maritime republics with specified trade routes to the Mediterranean countries and beyond. Pisa had played a powerful role in commerce since Roman times and even before as a port of call for Greek traders. Early on it had established outposts for its commerce along its colonies and trading routes.

Figure 1-1
Leonardo of Pisa (Fibonacci).

In 1192 Guilielmo Bonacci became a public clerk in the customs house for the Republic of Pisa, which was stationed in the Pisan colony of Bugia (later Bougie, and today Bejaia, Algeria) on the Barbary Coast of Africa. Shortly after his arrival he brought his son, Leonardo, to join him so that the boy could learn the skill of calculating and become a merchant. This ability was significant since each republic had its own units of money and traders had to calculate monies due them. This meant determining currency equivalents on a daily basis. It was in Bugia that Fibonacci first became acquainted with the "nine Indian figures," as he called the Hindu numerals and "the sign 0 which the Arabs call zephyr." He declares his fascination for the methods of calculation using these numerals in the only source we have about his life story, the prologue to his most famous book, *Liber Abaci*. During his time away from Pisa, he received instruction from a Muslim teacher, who introduced him to a book on algebra titled *Hisâb al-jabr w'al-muqabâlah*[2] by the Persian mathematician al-Khowarizmi (ca. 780–ca. 850), which also influenced him.

During his lifetime, Fibonacci traveled extensively to Egypt, Syria, Greece, Sicily, and Provence, where he not only conducted business but also met with mathematicians to learn their ways of doing mathematics. Indeed Fibonacci sometimes referred to himself as "Bigollo," which could mean good-for-nothing or, more positively, traveler. He may have liked the double meaning. When he returned to Pisa around the turn of the century, Fibonacci began to write about calculation methods with the Indian numerals for commercial applications in his book, *Liber Abaci*.[3] The volume is mostly loaded with algebraic problems and those "real-world" problems that require more abstract mathematics. Fibonacci wanted to spread these newfound techniques to his countrymen.

Bear in mind that during these times, there was no printing press, so books had to be written by the hands of scribes, and if a copy was to be made, that, too, had to be hand written. Therefore we are fortunate to still have copies of *Liber Abaci*, which first appeared in 1202 and was later revised in 1228.[4] Among Fibonacci's other books is *Practica geometriae* (1220), a book on the practice of geometry. It covers geometry and trigonometry with a rigor comparable to that of Euclid's work, and with ideas presented in proof form as well as in numerical form, using these very convenient numerals.

---

2. The name "algebra" comes from the title of this book.
3. The title means "book on calculation"; it is not about the abacus.
4. An English-language version of *Liber Abaci*—the first in a modern language—of the 1228, Latin version of 1857 by Baldassarre Boncompagni was recently published by Laurence E. Sigler (New York: Springer-Verlag, 2002).

Here Fibonacci uses algebraic methods to solve geometric problems as well as the reverse. In 1225 he wrote *Flos* (on flowers or blossoms) and *Liber quadratorum* (or "Book of Squares"), a book that truly distinguished Fibonacci as a talented mathematician, ranking very high among number theorists. Fibonacci likely wrote additional works that are lost. His book on commercial arithmetic, *Di minor guisa*, is lost, as is his commentary on Book X of Euclid's *Elements*, which contained a numerical treatment of irrational numbers[5] as compared to Euclid's geometrical treatment.

The confluence of politics and scholarship brought Fibonacci into contact with the Holy Roman Emperor Frederick II (1194–1250) in the third decade of the century. Frederick II, who had been crowned king of Sicily in 1198, king of Germany in 1212, and then crowned Holy Roman Emperor by the pope in St. Peter's Cathedral in Rome (1220), had spent the years up to 1227 consolidating his power in Italy. He supported Pisa, then with a population of about ten thousand, in its conflicts with Genoa at sea and with Lucca and Florence on land. As a strong patron of science and the arts, Frederick II became aware of Fibonacci's work through the scholars at his court, who had corresponded with Fibonacci since his return to Pisa around 1200. These scholars included Michael Scotus (ca. 1175–ca. 1236), who was the court astrologer and to whom Fibonacci dedicated his book *Liber Abaci*; Theodorus Physicus, the court philosopher; and Dominicus Hispanus, who suggested to Frederick II that he meet Fibonacci when Frederick's court met in Pisa around 1225. The meeting took place as expected within the year.

Johannes of Palermo, another member of Frederick II's court, presented a number of problems as challenges to the great mathematician Fibonacci. Fibonacci solved three of these problems. He provided solutions in *Flos*, which he sent to Frederick II. One of these problems, taken from Omar Khayyam's (1048–1122) book on algebra, was to solve the equation: $x^3 + 2x^2 + 10x = 20$. Fibonacci knew that this was not solvable with the numerical system then in place—the Roman numerals. He provided an approximate answer,

---

5. Irrational numbers are those that cannot be expressed as a common fraction involving integers.

since he pointed out that the answer was not an integer, nor a fraction, nor the square root of a fraction. Without any explanation, he gives the approximate solution in the form of a sexagesimal[6] number as 1.22.7.42.33.4.40, which equals:

$$1 + \frac{22}{60} + \frac{7}{60^2} + \frac{42}{60^3} + \frac{33}{60^4} + \frac{4}{60^5} + \frac{40}{60^6}$$

However, with today's computer algebra system we can get the proper (real) solution—by no means trivial! It is

$$x = -\sqrt[3]{\frac{2\sqrt{3,930}}{9} - \frac{352}{27}} + \sqrt[3]{\frac{2\sqrt{3,930}}{9} + \frac{352}{27}} - \frac{2}{3}$$
$$\approx 1.3688081078$$

Another of the problems with which he was challenged is one we can explore here, since it doesn't require anything more than some elementary algebra. Remember that although these methods may seem elementary to us, they were hardly known at the time of Fibonacci, and so this was considered a real challenge. The problem was to find the perfect square[7] that remains a perfect square when increased or decreased by 5.

Fibonacci found the number $\left(\frac{41}{12}\right)^2$ as his solution to the problem. To check this, we must add and subtract 5 and see if the result is still a perfect square.

$$\left(\frac{41}{12}\right)^2 + 5 = \frac{1,681}{144} + \frac{720}{144} = \frac{2,401}{144} = \left(\frac{49}{12}\right)^2$$

$$\left(\frac{41}{12}\right)^2 - 5 = \frac{1,681}{144} - \frac{720}{144} = \frac{961}{144} = \left(\frac{31}{12}\right)^2$$

We have then shown that $\frac{41}{12}$ meets the criteria set out in the problem. Luckily the problem asked for 5 to be added and subtracted from the perfect square, for if he were asked to add or sub-

---

6. A number in base 60.
7. A perfect square is the square of a number, e.g., 16, 36, or 81.

tract 1, 2, 3, or 4 instead of 5, the problem could not have been solved. For a more general solution of this problem, see appendix B.

The third problem, also presented in *Flos*, is to solve the following:

> Three people are to share an amount of money in the following parts: $\frac{1}{2}$, $\frac{1}{3}$, and $\frac{1}{6}$. Each person takes some money from this amount of money until there is nothing left. The first person then returns $\frac{1}{2}$ of what he took. The second person then returns $\frac{1}{3}$ of what he took, and the third person returns $\frac{1}{6}$ of what he took. When the total of what was returned is divided equally among the three, each has his correct share, namely, $\frac{1}{2}$, $\frac{1}{3}$, and $\frac{1}{6}$. How much money was in the original amount and how much did each person get from the original amount of money?

Although none of Fibonacci's competitors could solve any of these three problems, he gave as an answer of 47 as the smallest amount, yet he claimed the problem was indeterminate.

The last mention of Fibonacci was in 1240, when he was honored with a lifetime salary by the Republic of Pisa for his service to the people, whom he advised on matters of accounting, often pro bono.

The Book *Liber Abaci*

Although Fibonacci wrote several books, the one we will focus on is *Liber Abaci*. This extensive volume is full of very interesting problems. Based on the arithmetic and algebra that Fibonacci had accumulated during his travels, it was widely copied and imitated, and, as noted, introduced into Europe the Hindu-Arabic place-valued decimal system along with the use of Hindu-Arabic numerals. The book was increasingly widely used for the better part of the next two centuries—a best seller!

He begins *Liber Abaci* with the following:

The nine Indian[8] figures are: 9 8 7 6 5 4 3 2 1.[9]

With these nine figures, and with the sign 0, which the Arabs call zephyr, any number whatsoever is written, as demonstrated below. A number is a sum of units, and through the addition of them the number increases by steps without end. First one composes those numbers, which are from one to ten. Second, from the tens are made those numbers, which are from ten up to one hundred. Third, from the hundreds are made those numbers, which are from one hundred up to one thousand . . . and thus by an unending sequence of steps, any number whatsoever is constructed by joining the preceding numbers. The first place in the writing of the numbers is at the right. The second follows the first to the left.

Despite their relative facility, these numerals were not widely accepted by merchants, who were suspicious of those who knew how to use them. They were simply afraid of being cheated. We can safely say that it took the same three hundred years for these numerals to catch on as it did for the completion of the Leaning Tower of Pisa.

Interestingly, *Liber Abaci* also contains simultaneous linear equations. Many of the problems that Fibonacci considers, however, were similar to those appearing in Arab sources. This does not detract from the value of the book, since it is the collection of solutions to these problems that makes the major contribution to our development of mathematics. As a matter of fact, a number of mathematical terms—common in today's usage—were first introduced in *Liber Abaci*. Fibonacci referred to "factus ex multiplicatione,"[10] and from this first sighting of the word, we speak of the "factors of a number" or the "factors of a multiplication." Another example of words whose introduction into the current mathematics vocabulary seems to stem from this famous book are the words "numerator" and "denominator."

---

8. Fibonacci used the term Indian figures for the Hindu-Arabic numerals.
9. It is assumed that Fibonacci wrote the numerals in order from right to left, since he took them from the Arabs, who write in this direction.
10. David Eugene Smith, *History of Mathematics*, vol. 2 (New York: Dover, 1958), p. 105.

The second section of *Liber Abaci* includes a large collection of problems aimed at merchants. They relate to the price of goods, how to convert between the various currencies in use in Mediterranean countries, how to calculate profit on transactions, and problems that had probably originated in China.

Fibonacci was aware of a merchant's desire to circumvent the church's ban on charging interest on loans. So he devised a way to hide the interest in a higher initial sum than the actual loan, and based his calculations on compound interest.

The third section of the book contains many problems, such as:

> A hound whose speed increases arithmetically chases a hare whose speed also increases arithmetically. How far do they travel before the hound catches the hare?

> A spider climbs so many feet up a wall each day and slips back a fixed number each night. How many days does it take him to climb the wall?

> Calculate the amount of money two people have after a certain amount changes hands and the proportional increase and decrease are given.

There are also problems involving perfect numbers,[11] problems involving the Chinese remainder theorem, and problems involving the sums of arithmetic and geometric series. Fibonacci treats numbers such as $\sqrt{10}$ in the fourth section, both with rational[12] approximations and with geometric constructions.

Some of the classical problems, which are considered recreational mathematics today, first appeared in the Western world in *Liber Abaci*. Yet the technique for solution was always the chief concern for introducing the problem. This book is of interest to us, not only because it was the first publication in Western culture to use the Hindu numerals to replace the clumsy Roman numerals, or because Fibonacci was the first to use a horizontal fraction bar, but

---

11. Perfect numbers are those where the sum of the number's proper factors is equal to the number itself. For example, the proper factors of 6 are 1, 2, and 3. The sum of these factors is $1 + 2 + 3 = 6$, and therefore 6 is a perfect number. The next larger perfect number is 28.
12. Rational numbers are those that can be expressed as common fractions, involving integers.

because it casually includes a recreational mathematics problem in chapter 12 that has made Fibonacci famous for posterity. This is the problem on the regeneration of rabbits.

The Rabbit Problem

Figure 1-2 shows how the problem was stated (with the marginal notes included):

Beginning 1   A certain man had one pair of rabbits together in a certain enclosed place, and one wishes to know how many are created from the pair in one year when it is the nature of them in a single month to bear another pair, and in the second month those born to bear also. Because the above written pair in the first month bore, you will double it; there will be two pairs in one month. One of these, namely the first, bears in the second month, and thus there are in the second month 3 pairs; of these in one month two are pregnant and in the third month 2 pairs of rabbits are born and thus there are 5 pairs in the month; in this month 3 pairs are pregnant and in the fourth month there are 8 pairs, of which 5 pairs bear another 5 pairs; these are added to the 8 pairs making 13 pairs in the fifth month; these 5 pairs that are born in this month do not mate in this month, but another 8 pairs are pregnant, and thus there are in the sixth month 21 pairs; to these are added the 13 pairs that are born in the seventh month; there will be 34 pairs in this month; to this are added the 21 pairs that are born in the eighth month; there will be 55 pairs in this month; to these are added the 34 pairs that are born in the ninth month; there will be 89 pairs in this month; to these are added again the 55 pairs that are born in the tenth month; there will be 144 pairs in this month; to these are added again the 89 pairs that are born in the eleventh month; there will be 233 pairs in this month. To these are still added the 144 pairs that are born in the last month; there will be 377 pairs and this many pairs are produced from the above-written pair in the mentioned place at the end of one year.

First 2

Second 3
Third 5

Fourth 8
Fifth 13

Sixth 21
Seventh 34

Eighth 55
Ninth 89

Tenth 144
Eleventh 233

Twelfth 377

You can indeed see in the margin how we operated, namely, that we added the first number to the second, namely the 1 to the 2, and the second to the third and the third to the fourth and the fourth to the fifth, and thus one after another until we added the tenth to the eleventh, namely the 144 to the 233, and we had the above-written sum of rabbits, namely 377 and thus you can in order find it for an unending number of months.

Figure 1-2

To see how this problem would look on a monthly basis, consider the chart in figure 1-3. If we assume that a pair of baby (B) rabbits matures in one month to become offspring-producing adults (A), then we can set up the following chart:

Month	Pairs	No. of Pairs of Adults (A)	No. of Pairs of Babies (B)	Total Pairs
Jan. 1		1	0	1
Feb. 1		1	1	2
Mar. 1		2	1	3
Apr. 1		3	2	5
May 1		5	3	8
June 1	A B A A B A B A A B A A B	8	5	13
July 1		13	8	21
Aug. 1		21	13	34
Sept. 1		34	21	55
Oct. 1		55	34	89
Nov. 1		89	55	144
Dec. 1		144	89	233
Jan. 1		233	144	377

Figure 1-3

This problem generated the sequence of numbers:

1, 1, 2, 3, 5, 8, 13, 21, 34, 55, 89, 144, 233, 377, . . . ,

which is known today as the *Fibonacci numbers*. At first glance there is nothing spectacular about these numbers beyond the relationship that would allow us to generate additional numbers of the sequence quite easily. We notice that every number in the sequence (after the first two) is the sum of the two preceding numbers.

The Fibonacci sequence can be written in a way that its recursive definition becomes clear: each number is the sum of the two preceding ones.

$$1$$
$$1$$
$$1 + 1 = 2$$
$$1 + 2 = 3$$
$$2 + 3 = 5$$
$$3 + 5 = 8$$
$$5 + 8 = 13$$
$$8 + 13 = 21$$
$$13 + 21 = 34$$
$$21 + 34 = 55$$
$$34 + 55 = 89$$
$$55 + 89 = 144$$
$$89 + 144 = 233$$
$$144 + 233 = 377$$
$$233 + 377 = 610$$
$$377 + 610 = 987$$
$$610 + 987 = 1,597. . .$$

The Fibonacci sequence is the oldest known (recursive) *recurrent* sequence. There is no evidence that Fibonacci knew of this relationship, but it is securely assumed that a man of his talents and insight knew the recursive[13] relationship. It took another four hundred years before this relationship appeared in print.

Introducing the Fibonacci Numbers

These numbers were not identified as anything special during the time Fibonacci wrote *Liber Abaci*. As a matter of fact, the famous German mathematician and astronomer Johannes Kepler (1571–1630) mentioned these numbers in a 1611 publication[14] when he wrote of the ratios: "as 5 is to 8, so is 8 to 13, so is 13 to 21 almost." Centuries passed and the numbers still went unnoticed. In the 1830s C. F. Schimper and A. Braun noticed that the numbers appeared as the number of spirals of bracts on a pinecone. In the mid 1800s the Fibonacci numbers began to capture the fascination of mathematicians. They took on their current name ("Fibonacci numbers") from François-Édouard-Anatole Lucas[15] (1842–1891), the French mathematician usually referred to as "Edouard Lucas," who later devised his own sequence by following the pattern set by Fibonacci. Lucas numbers form a sequence of numbers much like the Fibonacci numbers and also closely related to the Fibonacci numbers. Instead of starting with 1, 1, 2, 3, . . . , Lucas thought of beginning with 1, 3, 4, 7, 11. . . . Had he started with 1, 2, 3, 5, 8, . . . , he would have ended up with a somewhat truncated version of

---

13. The recursive relationship comes from the notion that each number is generated from its two predecessors. This will be explored in more detail later in the book—in chapter 9.

14. Maxey Brooke, "Fibonacci Numbers and Their History through 1900," *Fibonacci Quarterly* 2 (April 1964):149.

15. Lucas is also well known for his invention of the Tower of Hanoi puzzle and other mathematical recreations. The Tower of Hanoi puzzle appeared in 1883 under the name of M. Claus. Notice that Claus is an anagram of Lucas.

His four-volume work on recreational mathematics (1882–1894) has become a classic. Lucas died as the result of a freak accident at a banquet when a plate was dropped and a piece flew up and cut his cheek. He died of erysipelas a few days later.

the Fibonacci sequence. (We will inspect the Lucas numbers later in this chapter.)

At about this time the French mathematician Jacques-Philippe-Marie Binet (1786–1856) developed a formula for finding any Fibonacci number given its position in the sequence. That is, with Binet's formula we can find the 118th Fibonacci number without having to list the previous 117 numbers. (We will have the opportunity to use this formula in chapter 9.)

Figure 1-4
François-Édouard-Anatole Lucas.[16]

Today, these numbers still hold the fascination of mathematicians around the world. The Fibonacci Association was created in the early 1960s to provide enthusiasts an opportunity to share ideas about these intriguing numbers and their applications. Through The *Fibonacci Quarterly*,[17] their official publication, many new facts, applications, and relationships about them can be shared worldwide. According to its official Web site, http://www.mscs .dal.ca/Fibonacci/, "The *Fibonacci Quarterly* is meant to serve as a focal point for interest in Fibonacci numbers and related questions, especially with respect to new results, research proposals, challenging problems, and innovative proofs of old ideas."

Still one may ask, what is so special about these numbers? That is what we hope to be able to demonstrate throughout this

---

16. Permission of Francis Lucas http://edouardlucas.free.fr/.
17. Gerald E. Bergum, editor, Department of Computer Science, South Dakota State University, Brookings, SD 57007.

book. Before we explore the vast variety of examples in which we find the Fibonacci numbers, let us begin by simply inspecting this famous Fibonacci number sequence and some of the remarkable properties it has.

We will use the symbol F_7 to represent the seventh Fibonacci number, and F_n to represent the nth Fibonacci number, or, as we say, the general Fibonacci number—that is, any Fibonacci number. Let us look at the first thirty Fibonacci numbers (figure 1-5). Notice that after we begin with the first two 1s, in every case after them, each number is the sum of the two preceding numbers.

$$F_1 = 1 \qquad\qquad F_{16} = 987$$
$$F_2 = 1 \qquad\qquad F_{17} = 1,597$$
$$F_3 = 2 \qquad\qquad F_{18} = 2,584$$
$$F_4 = 3 \qquad\qquad F_{19} = 4,181$$
$$\qquad\qquad\qquad\quad F_{20} = 6,765$$
$$F_5 = 5 \qquad\qquad F_{21} = 10,946$$
$$F_6 = 8 \qquad\qquad F_{22} = 17,711$$
$$F_7 = 13 \qquad\qquad F_{23} = 28,657$$
$$F_8 = 21 \qquad\qquad F_{24} = 46,368$$
$$F_9 = 34 \qquad\qquad F_{25} = 75,025$$
$$F_{10} = 55 \qquad\qquad F_{26} = 121,393$$
$$F_{11} = 89 \qquad\qquad F_{27} = 196,418$$
$$F_{12} = 144 \qquad\qquad F_{28} = 317,811$$
$$F_{13} = 233 \qquad\qquad F_{29} = 514,229$$
$$F_{14} = 377 \qquad\qquad F_{30} = 832,040$$
$$F_{15} = 610$$

Figure 1-5

In appendix A we provide you with a list of the first five hundred Fibonacci numbers. You may discover certain wonders there. For example, if you look at the list, you will notice that (as you would expect) the number of digits of the Fibonacci numbers increases steadily as the numbers get larger. The French mathematician Gabriel Lamé (1795–1870) proved that in the Fibonacci sequence there must be at least four numbers and at most five numbers with the same number of digits. In other words, there would never be only three Fibonacci numbers with a certain number of digits, nor would there be as many as six Fibonacci numbers with a specified number of digits. Interestingly, Lamé did not use the name "Fibonacci numbers." So, as a consequence, they were often called "Lamé numbers" because of his proof.

Another oddity that is easily noticeable by a quick inspection of the Fibonacci numbers is seen if you look at the sixtieth through the seventieth Fibonacci numbers, which we call F_{60} and F_{70}. You will notice a curious pattern emerging with their last digits. Yes, they form the Fibonacci sequence (last digits) starting from the beginning: 0, 1, 1, 2, 3, 5, 8, 13, 21, 34, 55.[18] This will continue on and on.

This form of a pattern of repetition, which mathematicians call "periodicity," can be seen most readily with rational numbers —those that can be written as common fractions. A fraction such as

$$\frac{1}{7} = 0.142857142857142857142857142857\ldots$$

which shows a continuous repetition of the digits 142857, can also be written as $\frac{1}{7} = 0.\overline{142857}$, with the bar above the digits representing the endless repetition of the digits it covers. Here the period is one involving six repeating digits. The Fibonacci numbers also exhibit a less obvious periodicity. Let's look at the first thirty-one Fibonacci numbers (plus the one preceding them, which we will call F_0). See figure 1-6. When F_0 through F_{31} are each divided by 7, we obtain the following sequence displaying the quotients and remainders.[19]

---

18. We can also begin the Fibonacci sequence with $F_0 = 0$ and $F_1 = 1$, instead of $F_1 = 1$ and $F_2 = 1$.
19. We are writing quotients in a multiplication form. That is, instead of writing, say, 55 divided by 7 is 7, with a remainder of 6, we will write this fact as $55 = 7 \cdot 7 + 6$.

$$F_0 = 0 = 0 \cdot 7 + 0 \qquad F_{16} = 987 = 141 \cdot 7 + 0$$

$$F_1 = 1 = 0 \cdot 7 + 1 \qquad F_{17} = 1,597 = 228 \cdot 7 + 1$$

$$F_2 = 1 = 0 \cdot 7 + 1 \qquad F_{18} = 2,584 = 369 \cdot 7 + 1$$

$$F_3 = 2 = 0 \cdot 7 + 2 \qquad F_{19} = 4,181 = 597 \cdot 7 + 2$$

$$F_4 = 3 = 0 \cdot 7 + 3 \qquad F_{20} = 6,765 = 966 \cdot 7 + 3$$

$$F_5 = 5 = 0 \cdot 7 + 5 \qquad F_{21} = 10,946 = 1,563 \cdot 7 + 5$$

$$F_6 = 8 = 1 \cdot 7 + 1 \qquad F_{22} = 17,711 = 2,530 \cdot 7 + 1$$

$$F_7 = 13 = 1 \cdot 7 + 6 \qquad F_{23} = 28,657 = 4,093 \cdot 7 + 6$$

$$F_8 = 21 = 3 \cdot 7 + 0 \qquad F_{24} = 46,368 = 6,624 \cdot 7 + 0$$

$$F_9 = 34 = 4 \cdot 7 + 6 \qquad F_{25} = 75,025 = 10,717 \cdot 7 + 6$$

$$F_{10} = 55 = 7 \cdot 7 + 6 \qquad F_{26} = 121,393 = 17,341 \cdot 7 + 6$$

$$F_{11} = 89 = 12 \cdot 7 + 5 \qquad F_{27} = 196,418 = 28,059 \cdot 7 + 5$$

$$F_{12} = 144 = 20 \cdot 7 + 4 \qquad F_{28} = 317,811 = 45,401 \cdot 7 + 4$$

$$F_{13} = 233 = 33 \cdot 7 + 2 \qquad F_{29} = 514,229 = 73,461 \cdot 7 + 2$$

$$F_{14} = 377 = 53 \cdot 7 + 6 \qquad F_{30} = 832,040 = 118,862 \cdot 7 + 6$$

$$F_{15} = 610 = 87 \cdot 7 + 1 \qquad F_{31} = 1,346,269 = 192,324 \cdot 7 + 1$$

Figure 1-6

You can see where we have 0 remainders; that is, where 7 divides the numbers, as with: F_0, F_8, F_{16}, and F_{24} in this list.

Notice the pattern of the remainders (since each F_n is the sum of the two preceding Fibonacci numbers):

$$0, 1, 1, 2, 3, 5, 1, 6, 0, 6, 6, 5, 4, 2, 6, 1$$

There is a sort-of-Fibonacci recursive relationship here, although a bit hidden. If you do the same recursive procedure as we used on the original Fibonacci numbers, but taking only the remainders when the number generated is divided by 7, we get the following. This is the pattern seen in figure 1-6.

0, 1, 1, 2, 3, 5, (8 = 7 + 1), **6,** (1 + 6 = 7 + **0**), 6, 6, (6 + 6 = 7 + **5**), (5 + 6 = 7 + **4**), (5 + 4 = 7 + **2**), 6, (2 + 6 = 7 + **1**), . . .

The bold numbers in the line above are precisely the numbers that we see as the sequence of remainders. That is, they form a recursive sequence much the same as the Fibonacci number sequence.

n	F_n
60	1548008755920
61	2504730781961
62	4052739537881
63	6557470319842
64	10610209857723
65	17167680177565
66	27777890035288
67	44945570212853
68	72723460248141
69	117669030460994
70	190392490709135

Figure 1-7

When you get to the 120th $\left(F_{120} \right)$, the same will be true (see figure 1-8). You might want to search in appendix A for other evidences of these Fibonacci numbers. And there will be other sightings! (Hint: look at F_{180}.)

n	F_n
120	5358359254990966640871840
121	8670007398507948658051921
122	14028366653498915298923761
123	22698374052006863956975682
124	36726740705505779255899443
125	59425114757512643212875125
126	96151855463018422468774568
127	155576970220531065681649693
128	251728825683549488150424261
129	407305795904080553832073954
130	659034621587630041982498215

Figure 1-8

Some Properties of the Fibonacci Numbers

To warm you up, we will present some of the countless curious characteristics of the Fibonacci numbers—working inductively. For those who care to see proofs of these relationships, we refer you to appendix B. So let's begin our inspection of these numbers.

1. The sum of any ten consecutive Fibonacci numbers is divisible by 11. We can convince ourselves that this may be true by considering some randomly chosen examples. Take, for example, the sum of the following ten consecutive Fibonacci numbers:

    $$13 + 21 + 34 + 55 + 89 + 144 + 233 + 377 + 610 + 987 = 2,563$$

    which just happens to be divisible by 11, since $11 \cdot 233 = 2,563$.

    Let's take another example: the sum of another ten consecutive Fibonacci numbers, say, from F_{21} to F_{30}:

    $$10,946 + 17,711 + 28,657 + 46,368 + 75,025 + 121,393$$
    $$+ 196,418 + 317,811 + 514,229 + 832,040 = 2,160,598$$

    which also is a multiple of 11, since $11 \cdot 196,418 = 2,160,598$.

    Are you convinced that the sum of any ten consecutive Fibonacci numbers is divisible by 11? You really shouldn't be convinced without further evidence. One way to go about persuading yourself of the truth in this "conjecture" is to keep on taking the sum of groups of ten consecutive Fibonacci numbers. You could also try to prove the statement, mathematically. (See appendix B.)

2. Two consecutive Fibonacci numbers do not have any common factors, which means that they are said to be relatively prime.[20] This can be seen by mere inspection. Or take any two consecutive Fibonacci numbers and factor them, and you will see that they have no common factors. Take a look at figure 1-9, where we have listed the first forty Fibonacci numbers

---

20. Two numbers (integers) are relatively prime if they have no common factors, other than 1.

and factored those that are not prime. Notice that no two consecutive Fibonacci numbers have any common factors. Enough examples of this should convince you. (However, for those who require a proof to be convinced that this is true for any pair of consecutive Fibonacci numbers, see appendix B.)

n	F_n	Factors
1	1	Unit
2	1	Unit
3	2	prime
4	3	Prime
5	5	prime
6	8	2^3
7	13	prime
8	21	$3 \cdot 7$
9	34	$2 \cdot 17$
10	55	$5 \cdot 11$
11	89	prime
12	144	$2^4 \cdot 3^2$
13	233	prime
14	377	$13 \cdot 29$
15	610	$2 \cdot 5 \cdot 61$
16	987	$3 \cdot 7 \cdot 47$
17	1,597	prime
18	2,584	$2^3 \cdot 17 \cdot 19$
19	4,181	$37 \cdot 113$
20	6,765	$3 \cdot 5 \cdot 11 \cdot 41$
21	10,946	$2 \cdot 13 \cdot 421$
22	17,711	$89 \cdot 199$
23	28,657	prime
24	46,368	$2^5 \cdot 3^2 \cdot 7 \cdot 23$
25	75,025	$5^2 \cdot 3,001$
26	121,393	$233 \cdot 521$
27	196,,418	$2 \cdot 17 \cdot 53 \cdot 109$
28	317811	$3 \cdot 13 \cdot 29 \cdot 281$
29	514,229	prime
30	832,040	$2^3 \cdot 5 \cdot 11 \cdot 31 \cdot 61$
31	1,346,269	$557 \cdot 2,417$
32	2,178,309	$3 \cdot 7 \cdot 47 \cdot 2,207$
33	3,524,578	$2 \cdot 89 \cdot 19,801$
34	5,702,887	$1,597 \cdot 3,571$
35	9,227,465	$5 \cdot 13 \cdot 141,961$
36	14,930,352	$2^4 \cdot 3^3 \cdot 17 \cdot 19 \cdot 107$
37	24,157,817	$73 \cdot 149 \cdot 2,221$
38	39,088,169	$37 \cdot 113 \cdot 9,349$
39	63,245,986	$2 \cdot 233 \cdot 135,721$
40	102,334,155	$3 \cdot 5 \cdot 7 \cdot 11 \cdot 41 \cdot 2,161$

Figure 1-9

3. Consider the Fibonacci numbers in the composite-number[21] positions (with the exception of the 4th Fibonacci number). By that we mean, the 6th, 8th, 9th, 10th, 12th, 14th, 15th, 16th, 18th, 20th, and so on, Fibonacci numbers—which are all nonprimes. A quick inspection (figure 1-10) shows that the Fibonacci numbers in the composite-number (i.e., nonprime) positions are also composite numbers. For the factors of these Fibonacci numbers, we refer you to figure 1-9 and for further evidence see the much longer list in appendix A. Again, you may be convinced of this property, but to really make it a fact, we have to prove it mathematically. (See appendix B.)

$F_6 = 8$ $F_{20} = 6,765$

$F_8 = 21$ $F_{21} = 10,946$

$F_9 = 34$ $F_{22} = 17,711$

$F_{10} = 55$ $F_{24} = 46,368$

$F_{12} = 144$ $F_{25} = 75,025$

$F_{14} = 377$ $F_{26} = 121,393$

$F_{15} = 610$ $F_{27} = 196,418$

$F_{16} = 987$ $F_{28} = 317,811$

$F_{18} = 2,584$ $F_{30} = 832,040$

Figure 1-10

One could conjecture at this point that the analog of this is also true—namely, that the Fibonacci numbers in prime-number[22] positions are also prime. That is, if we look at the list of the first thirty Fibonacci numbers above, those in the prime positions, namely, the 2nd, 3rd, 5th, 7th, 11th, 13th, 17th, 19th, 23rd, and 29th Fibonacci numbers, would then also have to be prime.

---

21. A composite number is one that is not prime, that is, it is divisible by numbers other than itself and 1.
22. A prime number is a number (other than 0 or ±1) that is divisible only by itself and 1.

They are:

$F_2 = 1$, not a prime

$F_3 = 2$

$F_5 = 5$

$F_7 = 13$

$F_{11} = 89$

$F_{13} = 233$

$F_{17} = 1,597$

$F_{19} = 4,181 = 37 \cdot 113$

$F_{23} = 28,657$

$F_{29} = 514,229$

However, you will notice that the 2nd and the 19th Fibonacci numbers are not primes, and therefore this analog situation does not hold true. Nothing further to prove. One counterexample suffices to draw this conclusion.

4. With all these lovely relationships embracing the Fibonacci numbers, there must be a simple way to get the sum of a specified number of these Fibonacci numbers. A simple formula would be helpful as opposed to actually adding all the Fibonacci numbers to a certain point. To derive such a formula for the sum of the first n Fibonacci numbers, we will use a nice little technique that will help us generate a formula.

Remember the basic rule (or definition of a Fibonacci number. We can write this definition formally as

$$F_{n+2} = F_{n+1} + F_n \text{, where } n \geq 1$$

Or

$$F_n = F_{n+2} - F_{n+1}$$

By substituting increasing values for n, we get:

$$F_1 = F_3 - F_2$$
$$F_2 = F_4 - F_3$$
$$F_3 = F_5 - F_4$$
$$F_4 = F_6 - F_5$$
$$\vdots$$
$$F_{n-1} = F_{n+1} - F_n$$
$$F_n = F_{n+2} - F_{n+1}$$

By adding these equations, you will notice that there will be many terms on the right side of the equations that will disappear (because their sum is zero—since you will be adding and subtracting the same number). What will remain on the right side will be $F_{n+2} - F_2 = F_{n+2} - 1$.

On the left side, we have the sum of the first n Fibonacci numbers: $F_1 + F_2 + F_3 + F_4 + \ldots + F_n$, which is what we are looking for.

Therefore, we get the following:

$$F_1 + F_2 + F_3 + F_4 + \ldots + F_n = F_{n+2} - 1$$

which says that the sum of the first n Fibonacci numbers is equal to the Fibonacci number two further along the sequence minus 1.

There is a shortcut notation that we can use to signify the sum: $F_1 + F_2 + F_3 + F_4 + \ldots + F_n$, and that is $\sum_{i=1}^{n} F_i$. This reads: "The sum of all the F_i terms where i takes on the values from 1 to n." So we can then write this result as:

$$\sum_{i=1}^{n} F_i = F_1 + F_2 + F_3 + F_4 + \ldots + F_n = F_{n+2} - 1$$

or simply: $\sum_{i=1}^{n} F_i = F_{n+2} - 1$

5. Suppose we now consider the sum of the consecutive even-positioned Fibonacci numbers, beginning with the first such Fibonacci number, F_2. Again, let's see if we can discover a pattern when adding these consecutive even-positioned Fibonacci numbers.

$$F_2 + F_4 = 1 + 3 = 4$$
$$F_2 + F_4 + F_6 = 1 + 3 + 8 = 12$$
$$F_2 + F_4 + F_6 + F_8 = 1 + 3 + 8 + 21 = 33$$
$$F_2 + F_4 + F_6 + F_8 + F_{10} = 1 + 3 + 8 + 21 + 55 = 88$$

We may notice that each sum is 1 less than a Fibonacci number—more particularly, the Fibonacci number that follows the last even number in the sum. We can write this symbolically as:

$$F_2 + F_4 + F_6 + F_8 + \ldots + F_{2n-2} + F_{2n} = F_{2n+1} - 1, \text{ where } n \geq 1$$

$$\text{or } \sum_{i=1}^{n} F_{2i} = F_{2n+1} - 1, \text{ where } n \geq 1$$

Again, this looks to be the case for all Fibonacci numbers, but remember we only looked at the first few. To accept this as general truth, we would have to prove it. (See appendix B.)

6. Now that we have established a short-cut method to get the sum of the initial consecutive even-positioned Fibonacci numbers, it is only fitting that we inspect the analog: the sum of the initial consecutive odd-positioned Fibonacci numbers. Once again, by considering a few examples of these sums, we will look for a pattern.

$$F_1 + F_3 = 1 + 2 = 3$$
$$F_1 + F_3 + F_5 = 1 + 2 + 5 = 8$$
$$F_1 + F_3 + F_5 + F_7 = 1 + 2 + 5 + 13 = 21$$
$$F_1 + F_3 + F_5 + F_7 + F_9 = 1 + 2 + 5 + 13 + 34 = 55$$

These sums appear to be Fibonacci numbers. But how do these sums relate to the series that generated them? In each case the sums are the next Fibonacci number after the last term in the sum of odd-positioned Fibonacci numbers.

This can be symbolically written as

$$F_1 + F_3 + F_5 + F_7 + \ldots + F_{2n-1} = F_{2n}$$

$$\text{or } \sum_{i=1}^{n} F_{2i-1} = F_{2n}$$

(See appendix B.)

If we add the sum of the initial consecutive even-positioned Fibonacci numbers and the sum of the initial consecutive odd-positioned Fibonacci numbers, we should get the sum of the initial consecutive Fibonacci numbers:

$$F_2 + F_4 + F_6 + F_8 + \ldots + F_{2n} = F_{2n+1} - 1, \text{ where } n \geq 1$$
$$F_1 + F_3 + F_5 + F_7 + \ldots + F_{2n-1} = F_{2n}, \text{where } n \geq 1$$

The sum of these sequences is:

$$F_1 + F_2 + F_3 + F_4 + \ldots + F_{2n} = F_{2n+1} - 1 + F_{2n}$$

or

$$F_1 + F_2 + F_3 + F_4 + \ldots + F_{2n} = F_{2n+2} - 1$$

which is consistent with what we concluded in item 4 (above).

7. Having established relationships for various sums of Fibonacci numbers, we shall now consider the sum of the squares of the initial Fibonacci numbers. Here we will see another astonishing relationship that continues to make the Fibonacci numbers special. Since we are talking about "squares," it is fitting for us to look at them geometrically.

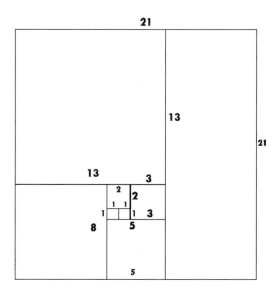

Figure 1-11

We find that, beginning with a 1×1 square (see figure 1-11), we can generate a series of squares, the sides of which are Fibonacci numbers. We can continue on this way indefinitely. Let us now express the area of each rectangle as the sum of its component squares.

$$1^2 + 1^2 + 2^2 + 3^2 + 5^2 + 8^2 + 13^2 = 13 \cdot 21$$

If we express the sum of the squares making up the smaller rectangles, we get this pattern:

$$1^2 + 1^2 = 1 \cdot 2$$
$$1^2 + 1^2 + 2^2 = 2 \cdot 3$$
$$1^2 + 1^2 + 2^2 + 3^2 = 3 \cdot 5$$
$$1^2 + 1^2 + 2^2 + 3^2 + 5^2 = 5 \cdot 8$$
$$1^2 + 1^2 + 2^2 + 3^2 + 5^2 + 8^2 = 8 \cdot 13$$
$$1^2 + 1^2 + 2^2 + 3^2 + 5^2 + 8^2 + 13^2 = 13 \cdot 21$$

From this pattern we can establish a rule: the sum of the squares of the Fibonacci numbers, to a certain point in the sequence, is equal to the product of the last number and the next number in the sequence. That is, if, for example, we choose to find the sum of the squares of the following portion of the sequence of Fibonacci numbers: 1, 1, 2, 3, 5, 8, 13, 21, 34, we would get:

$$1^2 + 1^2 + 2^2 + 3^2 + 5^2 + 8^2 + 13^2 + 21^2 + 34^2 = 1,870$$

However, we can apply this amazing rule that tells us that this sum is merely the product of the last number that is squared and the one that would come after it (sometimes known as its immediate successor), in the Fibonacci sequence. That would mean that this sum can be found by multiplying 34 by 55. Indeed, this product gives us $1,870 \, (= 34 \cdot 55)$.

We can write this with our summation notation as

$$\sum_{i=1}^{n} F_i^2 = F_n F_{n+1}$$

Imagine you would want to find the sum of the squares of the first thirty Fibonacci numbers. This neat relationship makes the task very simple. Instead of finding the squares of each and then finding their sum—a rather laborious task —all we need to do is multiply the thirtieth by the thirty-first Fibonacci number.

That is, the sum of the squares:

$$\sum_{i=1}^{30} F_i^2 = F_1^2 + F_2^2 + \ldots + F_{29}^2 + F_{30}^2$$

$$= 1^2 + 1^2 + 2^2 + 3^2 + 5^2 + 8^2 + 13^2 + \ldots + 514,229^2$$
$$+ \; 832,040^2$$
$$= 1,120,149,658,760$$

Or simply, using our newly established formula,

$$\sum_{i=1}^{30} F_i^2 = F_{30} \cdot F_{31} = 832,040 \cdot 1,346,269 = 1,120,149,658,760$$

(For the more advanced reader we offer a proof of this relationship in appendix B.)

8.   While we are talking about the squares of Fibonacci numbers, consider the following relationship. Let's take the square of a Fibonacci number, say, 34, and the square of the Fibonacci number that is two before it, 13. Subtracting these two squares, we get: $34^2 - 13^2 = 1{,}156 - 169 = 987$, which is also a Fibonacci number! So we squared the ninth Fibonacci number ($F_9 = 34$), and subtracted the square of the seventh Fibonacci number ($F_7 = 13$) to get the sixteenth Fibonacci number $\left(F_{16} = 987 \right)$. This can be written symbolically as: $F_9^{\,2} - F_7^{\,2} = F_{16}$.

It is nice to see that by subtracting the squares of two alternating[23] Fibonacci numbers, we seem to get another Fibonacci number. Will this work for any pair of alternating Fibonacci numbers? To answer this we would have to do a mathematical proof (which is provided in appendix B). However, we can begin to convince ourselves by inspecting some examples that this result was no fluke, but rather, that it actually works for any appropriate pair of Fibonacci numbers. Let's look at a few now and see if we can determine a pattern:

$$F_6^{\,2} - F_4^{\,2} = 8^2 - 3^2 = 55 = F_{10}$$
$$F_7^{\,2} - F_5^{\,2} = 13^2 - 5^2 = 144 = F_{12}$$
$$F_{15}^{\,2} - F_{13}^{\,2} = 610^2 - 233^2 = 317{,}811 = F_{28}$$

By inspecting the three subscripts for each of these examples, we notice that the sum of the first two equals the third. This would lead us to make the following generalization: $F_n^{2} - F_{n-2}^{2} = F_{2n-2}$. You might want to try to apply this

---

23. Alternating members of a sequence will be those that are one member apart from one another. That is, for example, the 4th and 6th members or the 15th and 17th members are called alternating members.

rule to a few more such pairs to further convince yourself if you wish. (A proof is in appendix B.)

9. The next natural question would be to look at the *sum* of the squares of two Fibonacci numbers. Suppose we consider two consecutive Fibonacci numbers, say, F_7 (=13) and F_8 (= 21). Their squares, 169 and 441, have a sum of 610, which is also a Fibonacci number, F_{15}. Perhaps we are onto something here. Let us try another consecutive pair of Fibonacci numbers and see what the sum of the squares of these numbers is. To establish a pattern, we will look at the two Fibonacci numbers F_{10} (= 55) and F_{11} (= 89). Their squares, 3,025 and 7,921, have a sum of 10,946, which is again a Fibonacci number, F_{21}. It appears that as we try this for several other pairs of consecutive Fibonacci numbers, we arrive each time at another Fibonacci number. But what is the pattern? To be able to predict which Fibonacci number will result from the sum of the squares of a consecutive pair of Fibonacci numbers, we will want to inspect their position in the sequence. In our first example, above, we used the Fibonacci numbers F_7 and F_8, and the sum of their squares was F_{15}. In the next example we used the Fibonacci numbers F_{10} and F_{11}, and we found that the sum of their squares was F_{21}. It appears that the sum of the subscripts of the two Fibonacci numbers whose squares we are adding will give us the subscript of the Fibonacci number representing their sum. That is, the sum of the squares of the Fibonacci numbers in positions n and $n + 1$ (consecutive positions) is the Fibonacci number in place $n + (n + 1)$ $= 2n + 1$, or $F_n^2 + F_{n+1}^2 = F_{2n+1}$. (The proof may be found in appendix B.)

10. Here is an engaging relationship that takes another look at the unexpected patterns we find among the numbers in the Fibonacci sequence. Take any four consecutive numbers in the sequence, say, 3, 5, 8, 13, and then find the difference of the squares of the middle two numbers: $8^2 - 5^2 = 64 - 25$ $= 39$. Then find the product of the outer two numbers:

$3 \cdot 13 = 39$. Amazingly the same result is attained! Was this by strange coincidence or was it a pattern that will hold for all such strings of four consecutive Fibonacci numbers? We can try this again with another group of four consecutive Fibonacci numbers, say, 8, 13, 21, 34. Again, we find the difference of the squares of the two middle numbers: $21^2 - 13^2 = 441 - 169 = 272$. If the pattern will continue to work, then the product of the outer two numbers must be 272. Well, $8 \cdot 34 = 272$, and so the pattern still holds. Symbolically, we can write for any four consecutive numbers, F_{n-1}, F_n, F_{n+1}, and F_{n+2}, this as $F_{n+1}^2 - F_n^2 = F_{n-1} \cdot F_{n+2}$. To really convince yourself that this is always true, it is not sufficient to merely try many groups of four consecutive Fibonacci numbers and find that the pattern holds for each of your tries, rather, you would have to do a proof. (See appendix B.)

11. Another curious relationship is obtained by inspecting the products of two alternating members of the Fibonacci sequence. Consider some of these products:

$F_3 \cdot F_5 = 2 \cdot 5 = 10$, which is 1 more than the square of the Fibonacci number 3: $F_4^2 + 1 = 3^2 + 1$.

$F_4 \cdot F_6 = 3 \cdot 8 = 24$, which is 1 less than the square of the Fibonacci number 5: $F_5^2 - 1 = 5^2 - 1$.

$F_5 \cdot F_7 = 5 \cdot 13 = 65$, which is 1 more than the square of the Fibonacci number 8: $F_6^2 + 1 = 8^2 + 1$.

$F_6 \cdot F_8 = 8 \cdot 21 = 168$, which is 1 less than the square of the Fibonacci number 13: $F_7^2 - 1 = 13^2 - 1$.

By now you should begin to see a pattern emerging: the product of two alternating Fibonacci numbers is equal to the square of the Fibonacci number between them ±1. We still need to determine when it is +1 or −1. When the number to be squared is in an even-numbered position, we add 1, and when it is in an odd-numbered position, we subtract 1. This can be generalized by using $(-1)^n$, since that will do exactly what we want: -1 to an even power is $+1$, and -1 to an odd power is -1.

Symbolically, this may be written as:

$$F_{n-1}F_{n+1} = F_n^2 + (-1)^n$$

where $n \geq 1$.

Although we "convinced" ourselves that this appears to be true, we must actually do a proof to be sure it is true for all cases. (See appendix B.)

This relationship can be expanded. Suppose instead of taking the product of the two Fibonacci numbers on either side of a particular Fibonacci number as we did above, we would take the two Fibonacci numbers that are one removed in either direction. Let's see how the product compares to the square of the Fibonacci number in the middle. If we take a specific example from the Fibonacci numbers, say, $F_6 = 8$, the product of the two Fibonacci numbers one removed on either side of 8 is $3 \cdot 21 = 63$, and the square of 8 is 64. They differ by 1. Suppose we now use $F_5 = 5$, then the product of the two numbers removed on either side is $2 \cdot 13 = 26$, which differs from 5^2 by 1. We can write this symbolically as:

$$F_{n-2}F_{n+2} = F_n^2 \pm 1$$

where $n \geq 1$.

You can now try this by comparing the product of Fibonacci numbers two, three, four, and so on removed on either side of a designated Fibonacci number, and you will find the following to hold true.

By now you may begin to recognize the pattern in the chart in figure 1-12. The difference between the square of the selected Fibonacci number and the various products of Fibonacci numbers, which are equidistant from the selected

Fibonacci number, is the square of another Fibonacci number. Symbolically we can write this as

$$F_{n-k}F_{n+k} - F_n^2 = \pm F_k^2$$

where $n \geq 1$ and $k \geq 1$.

Number (k) of Fibonacci numbers removed on either side of selected Fibonacci number	Symbolic Representation for the case of F_n		Example for $F_7 = 13$		Example for $F_8 = 21$		Difference between: $F_{n-k} \cdot F_{n+k}$ and F_n^2
1	$F_{n-1}F_{n+1}$	F_n^2	$8 \cdot 21 = 168$	$13^2 = 169$	$13 \cdot 34 = 442$	$21^2 = 441$	± 1
2	$F_{n-2}F_{n+2}$	F_n^2	$5 \cdot 34 = 170$	$13^2 = 169$	$8 \cdot 55 = 440$	$21^2 = 441$	± 1
3	$F_{n-3}F_{n+3}$	F_n^2	$3 \cdot 55 = 165$	$13^2 = 169$	$5 \cdot 89 = 445$	$21^2 = 441$	± 4
4	$F_{n-4}F_{n+4}$	F_n^2	$2 \cdot 89 = 178$	$13^2 = 169$	$3 \cdot 144 = 432$	$21^2 = 441$	± 9
5	$F_{n-5}F_{n+5}$	F_n^2	$1 \cdot 144 = 144$	$13^2 = 169$	$2 \cdot 233 = 466$	$21^2 = 441$	± 25
6	$F_{n-6}F_{n+6}$	F_n^2	$1 \cdot 233 = 233$	$13^2 = 169$	$1 \cdot 377 = 377$	$21^2 = 441$	± 64
k	$F_{n-k}F_{n+k}$	F_n^2					$\pm F_k^2$

Figure 1-12

There are endless properties that one can observe with the Fibonacci numbers. Many of them can be found by simple inspection. Before we embark on a more structured study of these amazing numbers, let's take another look at the sequence (figure 1-13).

F_1	1	F_{16}	897
F_2	1	F_{17}	1,597
F_3	2	F_{18}	2,584
F_4	3	F_{19}	4,181
F_5	5	F_{20}	6,765
F_6	8	F_{21}	10,946
F_7	13	F_{22}	17,711
F_8	21	F_{23}	28,657
F_9	34	F_{24}	46,368
F_{10}	55	F_{25}	75,025
F_{11}	89	F_{26}	121,393
F_{12}	144	F_{27}	196,418
F_{13}	233	F_{28}	317,811
F_{14}	377	F_{29}	514,229
F_{15}	610	F_{30}	832,040

Figure 1-13

12. Beginning with the first Fibonacci number, notice that every third number is even. That is, F_3, F_6, F_9, F_{12}, F_{15}, and F_{18} are all even numbers. We could restate this by saying that these numbers are all divisible by 2 or by F_3.

Let's look further at this list of Fibonacci numbers. Notice that every fourth number is divisible by 3. Here, F_4, F_8, F_{12}, F_{16}, F_{20}, and F_{24} are each divisible[24] by 3. We can again restate this by saying that F_4, F_8, F_{12}, F_{16}, F_{20}, and F_{24} are each divisible by F_4, or by 3.

Using patterns to see where we go from here, we would see that the first divisibility was by 2, then by 3, then, in good Fibonacci style, we ought to try to check for divisibility by 5—the next number in the Fibonacci sequence. Searching for Fibonacci numbers that are divisible by 5 is easy. They are the following numbers: 5, 55, 610, 6,765, 75,025 and 832,040, which correspond (symbolically) to F_5, F_{10}, F_{15}, F_{20},

---

24. You can check this quite easily by using the popular rule for divisibility by 3. If and only if the sum of the digits of a number is divisible by 3, then the number itself is divisible by 3.

F_{25}, and F_{30}. Thus, we are able to say that F_5, F_{10}, F_{15}, F_{20}, F_{25}, and F_{30} are each divisible by 5 or by F_5.

Checking for divisibility by 8 (the next Fibonacci number), we find that F_6, F_{12}, F_{18}, F_{24}, and F_{30} are each divisible by 8, or by F_6.

Yes, every seventh Fibonacci number is divisible by 13, or F_7. You might now try to generalize this finding. You can either say that a Fibonacci number F_{nm} is divisible by a Fibonacci number F_m, where n is a positive integer, or you may state it as the following: If p is divisible by q, then F_p is divisible by F_q. (A proof of this wonderful relationship is in appendix B.)

13. Fibonacci relationships, as noted earlier, can also be seen geometrically. We can consider Fibonacci squares as arranged as in figure 1-14 and figure 1-15.

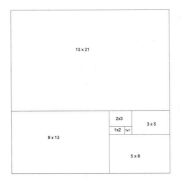

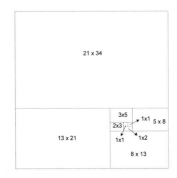

Figure 1-14
Fibonacci square with an odd number of rectangles, $n = 7$.

Figure 1-15
Fibonacci square with an odd number of rectangles, $n = 9$.

Figure 1-14 shows an odd number ($n = 7$) of rectangles into which the Fibonacci square is divided. The rectangles have dimensions of F_i and F_{i+1}, where i has values from 1 to n. The square then has four sides of length F_{n+1}. The area of the square is equal to the sum of the rectangles, which can constitute a geometric proof. This can be written symbolically as:

$$\sum_{i=2}^{n+1} F_i F_{i-1} = F_{n+1}^2$$

when n is odd.

As you can see in figure 1-14, when $n = 7$ the sum of the areas of the rectangles is

$$F_2 F_1 + F_3 F_2 + F_4 F_3 + F_5 F_4 + F_6 F_5 + F_7 F_6 + F_8 F_7$$
$$= 1 + 2 + 6 + 15 + 40 + 104 + 273 = 441 = 21^2 = F_8^2$$

On the other hand, when n is even (in figure 1-15, $n = 8$), that is, when a Fibonacci square is constructed using an even number of Fibonacci rectangles, a unit square remains that is not used. This indicates that the Fibonacci square is 1 unit larger than the sum of the rectangles, which gives us the following adjusted relationship:

$$\sum_{i=2}^{n+1} F_i F_{i-1} = F_{n+1}^2 - 1$$

when n is even.

In figure 1-15, where $n = 9$, the sum of the rectangles is

$$1 + F_2 F_1 + F_3 F_2 + F_4 F_3 + F_5 F_4 + F_6 F_5 + F_7 F_6 + F_8 F_7 + F_9 F_8$$
$$= 1 + 1 + 2 + 6 + 15 + 40 + 104 + 273 + 714$$
$$= 1{,}156 = 34^2 = F_9^2$$

14. Earlier in this chapter, we mentioned the Lucas numbers. They are: 1, 3, 4, 7, 11, 18, 29, 47, ... We can find the sum of the Lucas numbers much the same way we found the sum of the Fibonacci numbers. Again, there must be a simple way through a formula to get the sum of a specified number of these Lucas numbers. To derive such a formula for the sum of the first n Lucas numbers, we will use a nice little technique that will help us generate the formula.

The basic rule (or definition) of a Lucas number is

$$L_{n+2} = L_{n+1} + L_n, \text{ where } n \geq 1$$
$$\text{or } L_n = L_{n+2} - L_{n+1}$$

By substituting increasing values for n, we get:

$$L_1 = L_3 - L_2$$
$$L_2 = L_4 - L_3$$
$$L_3 = L_5 - L_4$$
$$L_4 = L_6 - L_5$$
$$\vdots$$
$$L_{n-1} = L_{n+1} - L_n$$
$$L_n = L_{n+2} - L_{n+1}$$

By adding these equations, you will notice that there will be many terms on the right side that will disappear because their sum is zero. This occurs since you will be adding and subtracting the same number. What will remain on the right side will be $L_{n+2} - L_2 = L_{n+2} - 3$.

On the left side, we have the sum of the first n Lucas numbers: $L_1 + L_2 + L_3 + L_4 + \ldots + L_n$, which is what we are looking for. Therefore, we get the following equation: $L_1 + L_2 + L_3 + L_4 + \ldots + L_n = L_{n+2} - 3$, which says that the sum of the first n Lucas numbers is equal to the Lucas number two further along the sequence minus 3.

There is a shortcut notation that we can use to signify the sum: $L_1 + L_2 + L_3 + L_4 + \ldots + L_n$, and that is $\sum_{i=1}^{n} L_i$. So we can then write this result as:

$$\sum_{i=1}^{n} L_i = L_1 + L_2 + L_3 + L_4 + \ldots + L_n = L_{n+2} - 3$$

or simply:

$$\sum_{i=1}^{n} L_i = L_{n+2} - 3$$

15. Just as we found the sum of the squares of the Fibonacci numbers, so, too, can we find the sum of the squares of the Lucas numbers.

Here we will uncover another astonishing relationship that continues to make the Lucas numbers special.

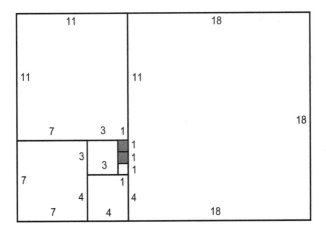

Figure 1-16

We find in figure 1-16 that, beginning with the three small 1×1 squares, we can generate a series of squares the sides of which are Lucas numbers. We can continue on this way indefinitely. Let us now express the area of the rectangle as the sum of its component squares minus the two small squares (shaded).

$$L_1^2 + L_2^2 + L_3^2 + L_4^2 + L_5^2 + L_6^2$$
$$= 1^2 + 3^2 + 4^2 + 7^2 + 11^2 + 18^2 = 520$$
$$= 522 - 2 = 18 \cdot 29 - 2$$
$$= L_6 \cdot L_7 - 2$$

If we express the sum of the squares progressively, which make up the smaller rectangles, we get this pattern:

$$1^2 + 3^2 = 3 \cdot 4 - 2$$
$$1^2 + 3^2 + 4^2 = 4 \cdot 7 - 2$$
$$1^2 + 3^2 + 4^2 + 7^2 = 7 \cdot 11 - 2$$
$$1^2 + 3^2 + 4^2 + 7^2 + 11^2 = 11 \cdot 18 - 2$$
$$1^2 + 3^2 + 4^2 + 7^2 + 11^2 + 18^2 = 18 \cdot 29 - 2$$
$$1^2 + 3^2 + 4^2 + 7^2 + 11^2 + 18^2 + 29^2 = 29 \cdot 47 - 2$$

From this pattern we can establish a rule: the sum of the squares of the Lucas numbers to a certain point in the sequence is equal to the product of the last number and the next number in the sequence minus 2. In other words, if, for example, we choose to find the sum of the squares of the following portion of the sequence of Lucas numbers: 1, 3, 4, 7, 11, 18, 29, 47, 76, we would get:

$$L_1^2 + L_2^2 + L_3^2 + L_4^2 + L_5^2 + L_6^2 + L_7^2 + L_8^2 + L_9^2$$
$$= 1^2 + 3^2 + 4^2 + 7^2 + 11^2 + 18^2 + 29^2 + 47^2 + 76^2$$
$$= 9{,}346$$

However, we can apply this amazing rule that tells us that this sum is merely the product of the last number and the one that would come after it (sometimes known as its immediate successor) in the Lucas sequence. That would mean that this sum can be found by multiplying 76 by 123 and subtracting 2. Indeed, this product gives us the following: $L_9 \cdot L_{10} = 76 \cdot 123 = 9{,}348 = 9{,}346 + 2$.

We can concisely summarize this as:

$$\sum_{i=1}^{n} L_i^2 = L_n L_{n+1} - 2$$

This will help you find the sum of the squares of the Lucas numbers without having to first find all the terms below it.

Imagine you would want to find the sum of the squares of the first thirty Lucas numbers. This neat relationship makes the task very simple. Instead of finding the squares of each and then finding their sum—a time consuming task—all you need to do is multiply the thirtieth by the thirty-first Lucas number and subtract 2. (See appendix B for a proof of this relationship.)

These proofs, although geometric, are reasonable enough to convince you of the veracity of these statements. Before having you explore more unexpected sightings of these lovely Lucas numbers, we provide you with a summary of the properties we have noted thus far.

A Summary of the Properties

Here is a summary of the Fibonacci (and Lucas) relationships we have seen (let be n any natural number; $n \geq 1$):

0. *Definition of the Fibonacci numbers F_n and of the Lucas numbers L_n:*

$$F_1 = 1; \ F_2 = 1; \ F_{n+2} = F_n + F_{n+1}$$
$$L_1 = 1; \ L_2 = 3; \ L_{n+2} = L_n + L_{n+1}$$

1. *The sum of any ten consecutive Fibonacci numbers is divisible by 11:*

$$11 \mid (F_n + F_{n+1} + F_{n+2} + \ldots + F_{n+8} + F_{n+9})$$

2. *Consecutive Fibonacci numbers are relatively prime: That is, their greatest common divisor is 1.*

3. *Fibonacci numbers in a composite number position (with the exception of the fourth Fibonacci number) are also composite numbers.*

Another way of saying this is that if n is not prime, then F_n is not prime

(with $n \neq 4$, since $F_4 = 3$, which is a prime number).

4. The sum of the first n Fibonacci numbers is equal to the Fibonacci number two further along the sequence minus 1:

$$\sum_{i=1}^{n} F_i = F_1 + F_2 + F_3 + F_4 + \ldots + F_n = F_{n+2} - 1$$

5. The sum of the consecutive even-positioned Fibonacci numbers is 1 less than the Fibonacci number that follows the last even number in the sum:

$$\sum_{i=1}^{n} F_{2i} = F_2 + F_4 + F_6 + \ldots + F_{2n-2} + F_{2n} = F_{2n+1} - 1$$

6. The sum of the consecutive odd-positioned Fibonacci numbers is equal to the Fibonacci number that follows the last odd number in the sum:

$$\sum_{i=1}^{n} F_{2i-1} = F_1 + F_3 + F_5 + \ldots + F_{2n-3} + F_{2n-1} = F_{2n}$$

7. The sum of the squares of the Fibonacci numbers is equal to the product of the last number and the next number in the Fibonacci sequence:

$$\sum_{i=1}^{n} F_i^2 = F_n F_{n+1}$$

8. The difference of the squares of two alternate[25] Fibonacci numbers is equal to the Fibonacci number in the sequence whose position number is the sum of their position numbers:

$$F_n^2 - F_{n-2}^2 = F_{2n-2}$$

---

25. I.e., a Fibonacci number with position number n and the Fibonacci number two before it.

9. *The sum of the squares of two consecutive Fibonacci numbers is equal to the Fibonacci number in the sequence whose position number is the sum of their position numbers:*

$$F_n^2 + F_{n+1}^2 = F_{2n+1}$$

10. *For any group of four consecutive Fibonacci numbers, the difference of the squares of the middle two numbers is equal to the product of the outer two numbers.*

Symbolically, we can write this as

$$F_{n+1}^2 - F_n^2 = F_{n-1} \cdot F_{n+2}$$

11. *The product of two alternating Fibonacci numbers is 1 more or less than the square of the Fibonacci number between them:*

$$F_{n-1}F_{n+1} = F_n^2 + (-1)^n$$

(If n is even, the product is 1 more; if n is odd, the product is 1 less.)

The difference between the square of the selected Fibonacci number and the various products of Fibonacci numbers equidistant from the selected Fibonacci number is the square of another Fibonacci number:

$$F_{n-k}F_{n+k} - F_n^2 = \pm F_k^2, \text{ where } n \geq 1, \text{ and } k \geq 1$$

12. *A Fibonacci number F_{mn} is divisible by a Fibonacci number F_m.*

We can write this as

$$F_m \mid F_{mn}, \text{ and it reads } F_m \text{ divides } F_{mn}$$

Or in other words: If p is divisible by q, then F_p is divisible by F_q or using our symbols:

$$q \mid p \Rightarrow F_q \mid F_p$$

(in the above, m, n, p, and q are positive integers)

Here is how this looks for specific cases:

$F_1 \mid F_n$,   i.e., $1 \mid F_1$, $1 \mid F_2$, $1 \mid F_3$, $1 \mid F_4$, $1 \mid F_5$, $1 \mid F_6$, ..., $1 \mid F_n$, ...
$F_2 \mid F_{2n}$,   i.e., $1 \mid F_2$, $1 \mid F_4$, $1 \mid F_6$, $1 \mid F_8$, $1 \mid F_{10}$, $1 \mid F_{12}$, ..., $1 \mid F_{2n}$, ...
$F_3 \mid F_{3n}$,   i.e., $2 \mid F_3$, $2 \mid F_6$, $2 \mid F_9$, $2 \mid F_{12}$, $2 \mid F_{15}$, $2 \mid F_{18}$, ..., $2 \mid F_{3n}$, ...
$F_4 \mid F_{4n}$,   i.e., $3 \mid F_4$, $3 \mid F_8$, $3 \mid F_{12}$, $3 \mid F_{16}$, $3 \mid F_{20}$, $3 \mid F_{24}$, ..., $3 \mid F_{4n}$, ...
$F_5 \mid F_{5n}$,   i.e., $5 \mid F_5$, $5 \mid F_{10}$, $5 \mid F_{15}$, $5 \mid F_{20}$, $5 \mid F_{25}$, $5 \mid F_{30}$, ..., $5 \mid F_{5n}$, ...
$F_6 \mid F_{6n}$,   i.e., $8 \mid F_6$, $8 \mid F_{12}$, $8 \mid F_{18}$, $8 \mid F_{24}$, $8 \mid F_{30}$, $8 \mid F_{36}$, ..., $8 \mid F_{6n}$, ...
$F_7 \mid F_{7n}$,   i.e., $13 \mid F_7$, $13 \mid F_{14}$, $13 \mid F_{21}$, $13 \mid F_{28}$, $13 \mid F_{35}$, $13 \mid F_{42}$, ..., $13 \mid F_{7n}$, ...

13. *The sum of the products of consecutive Fibonacci numbers is either the square of a Fibonacci number or one less than the square of a Fibonacci number.*

$$\sum_{i=2}^{n+1} F_i F_{i-1} = F_{n+1}^2, \text{ when } n \text{ is odd}$$

$$\sum_{i=2}^{n+1} F_i F_{i-1} = F_{n+1}^2 - 1, \text{ when } n \text{ is even}$$

14. *The sum of the first n Lucas numbers is equal to the Lucas number two further along the sequence minus 3:*

$$\sum_{i=1}^{n} L_i = L_1 + L_2 + L_3 + L_4 + \ldots + L_n = L_{n+2} - 3$$

15. *The sum of the squares of the Lucas numbers is equal to the product of the last number and the next number in the Lucas number sequence minus 2:*

$$\sum_{i=1}^{n} L_i^2 = L_n L_{n+1} - 2$$

Although our focus is largely about the Fibonacci numbers, we should not think of Fibonacci as a mathematician who is known only for this now-famous sequence of numbers that bears his name. Fibonacci, as mentioned, was one of the greatest mathematical influences in Western culture. He not only provided us with the tools to do efficient mathematics (e.g., the numerals heretofore known largely to the Eastern world), but he also introduced to the

European world a thought process that opened the way for many future mathematical endeavors.

Notice how the origin of the Fibonacci numbers, innocently embedded in a problem about the regeneration of rabbits, seems to have properties far beyond what might be expected. The surprising relationships of this sequence, within the realm of the numbers, are truly mind-boggling! It is this phenomenon coupled with the almost endless applications beyond the sequence that has intrigued mathematicians for generations. It is our intention to fascinate with applications of these numbers as far afield as one can imagine.

Figure 1-17
A statue of Fibonacci in Pisa.

Chapter 2

The Fibonacci Numbers in Nature

Fibonacci numbers, as we have seen briefly, have amazing properties. We will continue to explore the wide variety of applications and relationships that these numbers hold. Here we will go back to the original manifestation of the Fibonacci numbers: in nature. We first saw them evolve in the regeneration of rabbits. Let's see if there are other living creatures whose regeneration might result in an application of the Fibonacci numbers.

The Male Bee's Family Tree

Of the more than thirty thousand species of bees, the most well known is probably the honeybee, which lives in a bee hive and has a family. So let us take a closer look at this type of bee. Curiously enough, an inspection of the family tree of the male bee will reveal our famous numbers. To examine the family tree closely, we first have to understand the peculiarities of the male bee. In a bee hive there are three types of bees: the male bee (called a drone), who does no work; the female bees (called worker bees), who do all the work; and the queen bee, who produces eggs to generate more bees. A male bee hatches from an unfertilized egg, which means it has only a mother and no father (but does have a grandfather), whereas a female bee hatches from a fertilized egg, thus requiring a mother (queen bee) and a father (one of the drones). The female bees end up as worker bees, unless they have taken some of the "royal jelly" that enables them to become queen bees and begin new colonies of bees elsewhere.

In the table (figure 2-1, where ♂ represents the male and ♀ represents the female) we begin with the male bee whose ancestors we are tracking. As we said, the male bee must come from an unfertilized egg, so only a female was needed to produce him. However, the egg-producing female must have had a mother and a father, so the third line has both a male and a female. It then continues with this pattern: the immediate ancestor of every male is a single female and the ancestor of a female is a male and a female. At the right, is a summary of the number of bees in each row. As you look down each column at the right, you will recognize the familiar Fibonacci numbers. Surprised? If not, you ought to be. This is clearly a surprising appearance of the Fibonacci numbers. Even the biology here is quite different from the rabbit's reproductory pattern, which generated the first recognition of the Fibonacci numbers.

Geneological Table of a Male Bee (Drone)

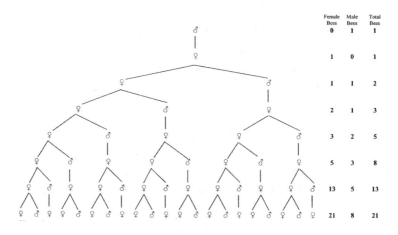

Female Bees	Male Bees	Total Bees
0	1	1
1	0	1
1	1	2
2	1	3
3	2	5
5	3	8
13	5	13
21	8	21

Figure 2-1

Had Fibonacci been aware of the relationship that unfolds as we inspect the genealogy of the drone (male bee), then he could have used it in place of what some consider as a less-than-realistic example of the regeneration of rabbits in his *Liber Abaci*. However, the rabbit problem is, in fact, not so far off from reality. After age three months, rabbits can produce offspring and can reproduce them on a monthly basis. So his rabbit problem was close to accurate.

The regeneration of the rabbits and the genealogy of the drone are described with the recursive formula $F_{n+2} = F_n + F_{n+1}$, which, as we have seen, means that each number in the sequence is the sum of its two immediate predecessors.

If we extend our theme here from the "Fibonacci numbers in nature" to the "Fibonacci numbers in society"—which could be stretched to be the nature of man!—then we can use the example of the spreading of rumors in society as an example of the Fibonacci numbers at work. We can create a model[1] for the spread of a rumor, which will generate the Fibonacci numbers.

The message (i.e., the rumor) will be passed on to only one person at a time. Specifically:

    (1) If person x is told the rumor, the earliest he will pass it on is the next day.
    (2) Person x passes the rumor to *only* one person per day.
    (3) After person x has passed the rumor twice, he is not interested in it any more, so person x does not go on spreading the rumor any further.

A chart of this stream of events should expose the Fibonacci sequence.

---

1. M. Huber, U. Manz, and H. Walser, *Annäherung an den Goldenen Schnitt* (Approaching the Golden Section) ETH Zürich, Bericht No. 93-01, 1993, p. 57.

Distribution of a Rumor

(x passes the rumor on to xa and xb)

Day 1
1

Day^2 2
1, $1a = 2$

Day 3
1, 2, $1b = 3$, $2a = 4$

Day 4
1, 2, 3, 4, $2b = 5$, $3a = 6$, $4a = 7$

Day 5
1, 2, 3, 4, 5, 6, 7, $3b = 8$, $4b = 9$, $5a = 10$, $6a = 11$, $7a = 12$

Day 6
1, 2, 3, 4, 5, 6, 7, 8, 9, 10, 11, 12, $5b = 13$, $6b = 14$, $7b = 15$,
$8a = 16$, $9a = 17$, $10a = 18$, $11a = 19$, $12a = 20$

Day 7
1, 2, 3, 4, 5, 6, 7, 8, 9, 10, 11, 12, 13, 14, 15, 16, 17, 18, 19, 20, $8b$
$= 21$, $9b = 22$, $10b = 23$, $11b = 24$, $12b = 25$, $13a = 26$, $14a = 27$,
$15a = 28$, $16a = 29$, $17a = 30$, $18a = 31$, $19a = 32$, $20a = 33$
...

Days n	1	2	3	4	5	6	7	8	9	10	11	12
Persons p	1	2	4	7	12	20	33	54	88	143	232	376
Difference	(1)	1	2	3	5	8	13	21	34	55	89	144

Figure 2-2

We find $p_n = F_{n+2} - 1$.

The rumor would be spread so much faster, if step (2) were to be changed to the following:

(2) At first person x passes on the news to p people, on the second day to q people.

---

2. The number after each of the equal signs indicates the running total of people with the rumor, so that the last number after the last equal sign is the total number for that day.

Fibonacci Numbers in the Plant World

Perhaps you would be most surprised to discover that these ubiquitous numbers also appear in the plant world. Take, for example, the pineapple. It might even be worth your while to get a pineapple and see for yourself that the Fibonacci numbers are represented on the pineapple. The hexagonal bracts on a pineapple can be seen to form three different direction spirals. In figures 2-3, 2-4, 2-5, and 2-6 you will notice that in the three directions there are 5, 8, and 13 spirals. These are three consecutive Fibonacci numbers.

Figure 2-3

Figure 2-4   **Figure 2-5**   **Figure 2-6**

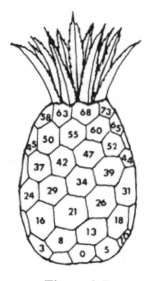

Figure 2-7

Consider the drawing of a pineapple (figure 2-7), where we have numbered the hexagons. The numbering has been done by the following rule: the lowest hexagon was assigned the number 0, the next higher one gets a 1 (remember this continues on the back side, not visible in the picture), then the next higher is assigned the number 2, and so on. Notice that hexagon 42 is slightly higher than hexagon 37. You should be able to identify three distinct spirals: one will have the hexagons 0, 5, 10, . . . ; the second will have hexagons 0, 13, 26, . . . ; and the third spiral would include hexagons 0, 8, 16, ... Now look at the common difference between the spirals and you will find the differences to be 5, 8, and 13—once again Fibonacci numbers.

The Pinecone

There are various species of pinecones (e.g., Norway spruce; Douglas fir or spruce; larch). Most will have two distinct direction spirals. Spiral arrangements are classified by the number of visible spirals (parastichies) that they exhibit. The number of spirals in each direction will be most often two successive Fibonacci numbers. When this occurs,[3] the angle between successive leaves or botanical elements is close to the "golden angle"—about 137.5°, which is related to the golden ratio (see chapter 4). The connection between the golden angle and the Fibonacci numbers can be seen in the ratio $\frac{137.5°}{360°}$, which becomes immediately clear when we reduce the fraction (see chapter 5) as:

---

3. In spiral phyllotaxis, botanical elements grow one by one, each at a constant divergence angle d from the previous one. This is the most common pattern, and most often the divergence angle d is close to the golden angle. The latter case gives rise to Fibonacci phyllotaxis.

$$\frac{137.5°}{360°} = \frac{55}{144}$$

The two pictures in figure 2-8 will bear this out, but you may want to convince yourself by getting some actual pinecones and counting the spirals yourself.

 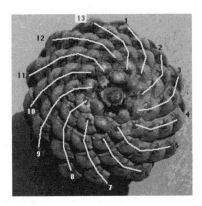

Figure 2-8

These pinecones have eight spirals in one direction and thirteen in the other direction. Again, you will notice these are Fibonacci numbers.

Tree (Species)	Number of spirals in one direction	Number of spirals in the other direction
Norway spruce	13	8
Douglas fir (spruce)	3	5
larch	5	3
pine	5	8

More details can be found in an article by Brother Alfred Brousseau.[4]

The following designations of spiral patterns do not make any pretense of completeness. Actually bracts can be lined up into sequences in many ways. The following are simply some of the more obvious patterns that have been observed. (As for notation, 8-5, for example, means that starting from a given bract and proceeding

---

4. Brother A. Brousseau, "On the Trail of the California Pine Spiral Patterns on California Pines," *Fibonacci Quarterly* 6, no. 1 (1968): 76.

along two spirals, 8 bracts will be found on one spiral and 5 on the other when going to the next intersection of the spirals.)

Pinus albicaulis (Whitebark pine),	5-3, 8-3, 8-5
Pinus flexilis (Limber pine),	8-5, 5-3, 8-3
Pinus Lamberttana (Sugar pine),	8-5, 13-5, 13-8, 3-5, 3-8, 3-13, 3-21
Pinus monticola (Western white pine, Silver pine),	3-5
Pinus monophylla (One-leaved pinon),	3-5, 3-8
Pinus edulis,	5-3
Pinus quadrifolia (Four-leaved pinon),	5-3
Pinus aristata (Bristlecone pine),	8-5, 5-3, 8-3
Pinus Balfouriana (Foxtail pine),	8-5, 5-3, 8-3
Pinus muricata (Bishop pine),	8-13, 5-8
Pinus remorata (Santa Cruz Island pine),	5-8
Pinus contorta (Beach pine),	8-13
Pinus Murrayana (Lodgepole pine, Tamarack pine),	8-5, 13-5, 13-8
Pinus Torreyana (Torrey pine),	8-5, 13-5
Pinus ponderosa (Yellow pine),	13-8, 13-5, 8-5
Pinus Jeffreyi (Jeffrey pine),	13-5, 13-8, 5-8
Pinus radiata (Monterey pine),	13-8, 8-5, 13-5
Pinus attenuata (Knobcone pine),	8-5, 13-5, 3-5, 3-8
Pinus Sabiniana (Digger pine),	13-8
Pinus Coulteri (Coulter pine),	13-8

In a similar detailed article we find spiral patterns, especially 5-8, in Aroids. The Aroids (family Araceae) are a group of attractive ornamental plants which include the very familiar *Aglaonema, Alocasia, Anthurium, Arum, Caladium, Colocasia, Dieffenbachia, Monstera, Philodendron, Scindapsus,* and *Spathiphyllum* species.[5] Spirals are also found on a variety of plants. Here are a few representative examples. You may well want to search for other examples yourself.

---

5. T. Antony Davis and T. K. Bose, "Fibonacci System in Aroids," *Fibonacci Quarterly* 9, no. 3 (1971): 253–55.

Figure 2-9
Mammilaria huitzilopochtli
has 13 and 21 spirals.

Figure 2-10
Mammilaria magnimamma
has 8 and 13 spirals.

Figure 2-11
Marguerite has 21 and
34 spirals.

Figure 2-12
Gymnocalcium izozogsii
has twice 5 and 8 spirals
(i.e., 10 and 16 spirals).

Figure 2-13
Knautia arvensis has twice
2 and 5 spirals (i.e., 4 and
10 spirals).

Figure 2-14
This anonium has 3
and 5 spirals.

Figure 2-15
Anonium.

Figure 2-16
Anonium.

The sunflower, on the other hand, has a variety of different spiral numbers. The older the flowers get, the more spirals they develop. In any case, however, the number of spirals will be a Fibonacci number. They usually will take on the following pairs of Fibonacci numbers: 13 (left-oriented spirals); 21 (right-oriented spirals), 21:34, 34:55, 55:89, 89:144.

Figure 2-17
Sunflower 1.

Figure 2-18
Sunflower 2.

Figure 2-19
Sunflower 3.

Figure 2-20
Sunflower 4.

In 2004 the principality of Liechtenstein[6] devoted an entire postage stamp series to science, called the "Exact Sciences; (CHF- .85) Mathematics; 2004." The stamp depicts the logarithmic spiral, exponential growth, and the 39th Mersenne prime number,[7]

$$2^{13,466,917} - 1$$

One of the subjects highlighted a sunflower.[8]

Figure 2-21

Based on a survey of the literature encompassing 650 species and 12,500 specimens, R. Jean[9] estimated that, among plants dis-

6. An independent state between Austria and Switzerland.
7. The 39th Mersenne prime number was found in 2001; and in 2005 we have come as far as finding the 42nd Mersenne prime number $2^{25\,964\,951} - 1$.
8. For sunflowers there exists three kinds of spirals: 82 percent are Fibonacci spirals (8, 13, 21, 34, . . .); 14 percent are Lucas spirals (7, 11, 18, 29, . . .); and 2 percent are Fibonacci bijugate spirals (10, 16, 26, 42, . . .). J. C. Schoute, "Early Binding Whorls," *Rec. Trav. Bot. Neer* 1, no. 35 (1938): 416–558.
9. Roger V. Jean, *Phyllotaxis: A Systemic Study in Plant Morphogenesis* (Cambridge: Cambridge University Press, 1994).

playing spiral or multijugate phyllotaxis,[10] about 92 percent of them have Fibonacci phyllotaxis.

One may wonder why this happens. Certainly mathematicians did not influence this phenomenon. The reason seems to be that this arrangement forms an optimal packing of the seeds so that, no matter how large the seed range, with all the seeds being the same size, they should be uniformly packed at any stage; this means no crowding in the center and not too sparse near the edges. We "see" this packing as spirals, which almost always translates into adjacent Fibonacci numbers.

Artificial Fibonacci-Flower Spirals

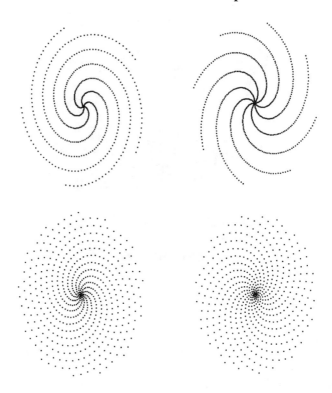

Figure 2-22

---

10. From the Greek: *Phyllo* means leaf, and *taxis* means arrangement.

Leaf Arrangement—Phyllotaxis

Having now concentrated at the center of the sunflower, or, for that matter, the daisy, we can now inspect the petals surrounding the center. Again, you will discover that most plants will have the number of such petals corresponding to a Fibonacci number. For example, lilies and irises have 3 petals; buttercups have 5 petals; some delphiniums have 8 petals; corn marigolds have 13 petals; some asters have 21 petals; and daisies can be found with 34, 55, or 89 petals.

Here is a short list with some flowers arranged by the number of petals they generally have:

3 petals: *iris*, snowdrop, lily (some lilies have 6 petals formed from two sets of 3)

5 petals: buttercup, columbine (*aquilegia*), wild rose, larkspur, pinks; also apple blossom, *hibiscus*

8 petals: *delphiniums* (larkspurs), *Cosmos bipinnatus*,[11] *Coreopsis tinctoria*

13 petals: corn marigold, *cineraria*, some daisies, ragwort

21 petals: aster, chicory, *Helianthus annuus*

34 petals: *pyrethrum* and other daisies

55, 89 petals: michaelmas daisies and other Compositae (*asteraceae* family)

Some species, such as the buttercup, are very precise about the number of petals they have, but others have petals whose numbers are very near those above—with the average being a Fibonacci number!

Having now inspected the parts of flowers, we can look at the placement of the leaves on a stem. Take a plant that has not been pruned and locate the lowest leaf. Begin with the bottom leaf, and count the number of rotations around the stem, each time going

---

11. These are species of Compositae—one of the largest families of the vascular plants. See P. P. Majumder and A. Chakravati, "Variation in the Number of Rays and Disc-Florets in Four Species of Compositae," *Fibonacci Quarterly* 14 (1976): 97–100.

through the next leaf up the stem, until you reach the next leaf whose direction is the same as the first leaf you identified (that is, above it and pointing in the same direction). The number of rotations will be a Fibonacci number. Furthermore, the number of leaves that you will pass along the way to reach the "final" leaf will also be a Fibonacci number.

In figure 2-23, it took five revolutions to reach the leaf (their eighth leaf) that is in the same direction as the first one. This phyllotaxis (i.e., leaf arrangement) will vary with different species, but should be a Fibonacci number. If we refer to the rotation/leaf number ratio for this plant, it would be about $\frac{5}{8}$. One also describes the curve marked in figure 2-23 as the "genetic spiral of a plant."

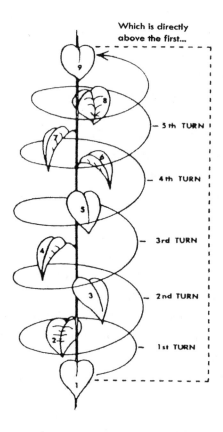

Figure 2-23

Here are some phyllotaxis ratios:

1/2: elm, linden, lime, some grasses
3/8: asters, cabbages, poplar, pear, hawkweed, some roses
1/3: alders, birches, sedges, beech, hazel, blackberry, some grasses
2/5: roses, oak, apricot, cherry, apple, holly, plum, common
    groundsel
8/21: fir trees, spruce
5/13: pussy willow, almond
13/34: some pine trees

Different species of palms display different numbers of leaf spirals, but the numbers always match with Fibonacci numbers. For example, in the arecanut palm (*Areca catechu*), or the ornamental *Ptychosperma macarthurii* palm, only a single foliar spiral is discernible, while in the sugar palm (*Arenga saccharifera*), or *Arenga pinnata*, two spirals each are visible. In the palmira palm (*Borassus flabellifer*), or *Corypha elata*, as well as in a number of other species of palms, three clear spirals are visible. The coconut palm (*Cocos nucifera*) as well as *Copernicia* spirals have five spirals, while the African oil palm (*Elaeis guineensis*) bears eight spirals. The wild date palm (*Phoenix sylvestris*) and a few other species of palms also show eight spirals. On stout trunks of the Canary island palm (*Phoenix canariensis*), thirteen spirals can be observed. Also in some of these plants, twenty-one spirals can be found. Palms bearing 4, 6, 7, 9, 10, or 12 obvious leaf spirals are not known.[12]

This phyllotaxis ratio of Fibonacci numbers is not guaranteed for every plant, but it can be observed on most plants. Why do these arrangements occur? We can speculate that some of the cases may be related to maximizing the space for each leaf, or the average amount of light falling on each one. Even a tiny advantage would come to dominate, over many generations. In the case of close-packed leaves in cabbages and succulents, such arrangements may be crucial for availability of space.

---

12. T. Antony Davis, "Why Fibonacci Sequence for Palm Leaf Spirals?" *Fibonacci Quarterly* 9 no. 3(1971): 237–44.

Much has been written about the arrangement of leaves, petals, and other plant aspects, but earlier works were just descriptive and did not explain any connection between numbers and plant growth. Rather they dealt with the geometry of the arrangements. The most dramatic insight comes out of a recently published work by the French mathematical physicists Stéphane Douady and Yves Couder. They developed a theory of the dynamics of plant growth using computer models and laboratory experiments in relation to the Fibonacci pattern. Douady and Couder[13] also found a dynamic explanation for the golden angle (137.5°). They obtained this angle as a consequence of simple rules of dynamics and did not postulate it as many of their predecessors, who saw it as a consequence of a space-saving arrangement. The spiral arrangement does not require any special botanical explanation.

Last, but not least, we can continue our search for Fibonacci numbers on the human body: A human being has 1 head, 2 arms, and 3 finger joints, and 5 fingers on each arm—these are all Fibonacci numbers. Wow! But, actually, it does not go on like this!

Moreover, the "small" Fibonacci numbers are just pure coincidence, because among the eight numbers from 1 to 8, there are five Fibonacci numbers. Therefore you have a good chance of finding a Fibonacci number purely by coincidence. The percentage of Fibonacci numbers among the rest of all other numbers changes dramatically if we look at bigger intervals of numbers. On the other hand, if you can find two different large Fibonacci numbers in one sunflower, you might ask, how this could be? (See page 68).

Particularly at the end of the nineteenth century, many scientists were of the opinion that the golden section was a divine, universal law of nature. Yet the golden section, ϕ, can be seen as a limit of a sequence, that is, made of the ratio of successive Fibonacci numbers. (This will be covered in chapter 4 in great detail.) So, how then can the human being, the "pride of creation," not have been designed by the rules of the golden section?! Some see it as if it were. Leonardo

---

13. Stéphane Douady and Yves Couder, "Phyllotaxis as a Physical Self-Organized Growth Process," *Physical Review Letters* 68, no. 13 (1992): 2098–101.

da Vinci (1452–1519)[14] constructed the proportion of the human body based on the golden section, ϕ (see pages 257–60). Furthermore, da Vinci partitioned the face vertically into thirds: forehead, middle face, and chin. Plastic surgeons further divide the face horizontally in fifths. In this way, we see once again the presence of the two Fibonacci numbers, 3 and 5, come into play.

The Belgian mathematician and astronomer and founder of the social statistics Lambert Adolphe Jacques Quetelet (1796–1874)[15] and the German author, critic, playwright, poet and philosopher Adolph Zeising (1810–1876)[16] measured the human body and saw its proportions related to the golden section, and thus greatly influenced future generations.

The French architect Le Corbusier (Charles-Edouard Jeanneret, 1887–1965) assumed that human body proportions are based on the golden section. Le Corbusier formulated the ideal proportions in the following way: height 182 cm; navel height 113 cm; fingertips with arms upraised 226 cm. The ratio of height to navel height is $\frac{182}{113} \approx 1.610619469$, a very close approximation of ϕ. By another measurement of the same features he got an analogous ratio of $\frac{176\,cm}{109\,cm} \approx 1.6147$, which is another good approximation of the ratio of two Fibonacci numbers, 13 and 21, or $\frac{21}{13} \approx 1.615384615$.

Looking further—maybe a bit too far!—the American Frank A. Lonc[17] measured sixty-five women to substantiate the work of da Vinci and Zeising by determining the mean ratio of a woman's height to her navel. He then established that the ideal proportions were those nearest 1.618, or approximately the golden section. In-

---

14. Italian painter, sculptor, architect, and engineer.
15. "Des proportions du corps humain," *Bulletin de l'Académie Royale des sciences, des lettres et des beaux-arts de Belgique*, vol. 1. Bruxelles 1848—15,1, pp. 580–93, and vol. 2.—15,2. pp. 16–27.
16. "Neue Lehre von den Proportionen des menschlichen Körpers" (Leipzig: R. Weigel, 1854).
17. Martin Gardner, "About Phi," *Scientific American*, August 1959, pp. 128–34.

deed T. Antony Davis and Rudolf Altevogt[18] claimed that the value of "a good-looking human body" based on Le Corbusier for both boys and girls of a similar age is approximately the golden ratio. Specifically, using 207 pupils in Münster, Germany, and 252 youngsters from Calcutta, India, they derived the value 1.615.

The ubiquitous appearances of the Fibonacci numbers in nature are just another piece of evidence that these numbers have truly phenomenal qualities.

---

18. "Golden Mean of the Human Body," *Fibonacci Quarterly* 17, no. 4 (1979): 340–44.

Chapter 3

The Fibonacci Numbers and the Pascal Triangle

B y now you probably expect the Fibonacci numbers to appear
in the most unexpected places. "Sightings" of the Fibonacci
numbers can either be of individual members of the sequence, as in
the case of many of their appearances in nature, or as a sequence of
numbers appearing where you least expect them. We will begin
with some unusual sequences and then bring the Fibonacci numbers into the discussion—yes, when you least expect them to appear. Not only will they appear, but they will interact with various
other fields in a very meaningful way.

Some Sequences

So far we have seen that the Fibonacci numbers emanate from a
nicely defined sequence, originally determined by a problem dealing with the regeneration of rabbits. We know they follow a
rule—that beginning with two 1s, the sum of every two consecutive members of the sequence gives the next number in the sequence. This may not be the usual kind of sequence to which we
are accustomed. We usually like sequences such as:

$$1, 2, 3, 4, 5, 6, 7, \ldots$$

or

$$2, 4, 6, 8, 10, 12, 14, \ldots$$

or

$$1, 3, 5, 7, 9, 11, 13, \ldots$$

We can even easily accept:

$$5, 10, 15, 20, 25, 30, 35, 40, \ldots$$

or even the sequence of perfect squares:

$$1, 4, 9, 16, 25, 36, \ldots$$

Suppose we didn't recognize that the above was the sequence of perfect squares; we could check the differences between terms to see if there is a common difference (or perhaps a common factor). For example, if there is a common difference, then you can easily get the next number, but if there is no common difference, then you could try to get the next row of differences—that is, between the terms of differences. Look at the chart below (figure 3-1) and notice that the second differences are constant.

Original Sequence	1		4		9		16		25		36
First Difference		3		5		7		9		11	
Second Difference			2		2		2		2		

Figure 3-1

However, when we get the sequence 1, 2, 4, 8, 16, we expect the next number to be 32. Yet if we were told that it is 31, we might cry out "wrong!"

Much to many people's surprise, either 32 or 31 "works" to create a viable or meaningful sequence.

Yes, what may be much to your amazement, **1, 2, 4, 8, 16, 31,** . . . can be a legitimate sequence.

To show that this is a proper sequence (one that has a rule for getting additional members of the sequence), we shall set up a table of differences. Just as we might have done for the first four sequences above, this will enable us to determine the rule for deriving succeeding members of the sequence.

We will set up a chart (figure 3-2) showing the differences between terms of the sequence **1, 2, 4, 8, 16, 31, . . .** , and then the differences of the differences, and so on until a pattern emerges. That is, we take the differences of the terms in the given sequence, and if no pattern emerges, we shall try to find a pattern in the new sequence formed by these differences. We will do this in exactly the same way: take differences between the terms and look for a pattern. The first row of differences leaves us without much of a clue of a pattern, since again the last term ruins what may have developed into a pattern. The second row, likewise. However, at the third sequence of differences, a clear pattern emerges: 1, 2, 3, ...

Original Sequence	1		2		4		8		16		31
First Difference		1		2		4		8		15	
Second Difference			1		2		4		7		
Third Difference				1		2		3			
Fourth difference					1		1				

Figure 3-2

With the fourth differences forming a sequence of constants, we can reverse the process (i.e., turn the table upside down as in figure 3-3) and extend the third differences a few more steps, say, up to 9.

Fourth Difference				1		1		1		1		1		1		1		1		1
Third Difference			1		2		3		4		5		6		7		8		9	
Second Difference		1		2		4		7		11		16		22		29		37		46
First Difference	1		2		4		8		15		26		42		64		93		130	
Original Sequence	1		2		4		8		16		31		57		99		163		256		386

Figure 3-3

The bold numbers are those that were obtained by working backward from the sequence of third differences. So the next numbers of the given sequence are 57, 99, 163, 256, and 386.[1]

After we have just taken you from what you expected would be a comfortable and familiar sequence, 1, 2, 4, 8, 16, 32, 64, 128, 256, . . . , to a rather strange-looking one, 1, 2, 4, 8, 16, 31, 57, 99, 163, 256, 386, . . . , you should not think that this sequence is contrived and is independent of other parts of mathematics. A geometric interpretation of this unusual sequence should help convince you of the consistency we enjoy in mathematics, and give you reason to better appreciate the beauty inherent in mathematics.

Consider a sequence of circles, each with successively more points highlighted on it. Beginning with the first circle, we will join these points with straight line segments to form the greatest number of regions in each of these circles. Let's count the number of regions in each circle successively. In figure 3-4 we show five circles, each with successively more points on it. We have omitted the actual "first circle"—the one with only one point on it, since it would show only one region formed—a trivial case. We marked the regions with consecutive numbers so that the number of regions can be easily counted.

---

1. The general term (that is, the nth term) of the sequence 1, 2, 4, 8, 16, 32, . . . is easy to find, and it is $T(n) = 2^{n-1}$. The general term of the sequence 1, 2, 4, 8, 16, 31, . . . is a fourth-power expression, since we had to go to the fourth differences to get a constant.

The general term is $T(n) = \dfrac{n^4 - 6n^3 + 23n^2 - 18n + 24}{24}$. [But also the general term $T(n) =$ $n + \binom{n}{4} + \binom{n-1}{2} = \binom{n}{4} + \binom{n}{2} + 1$ is for all natural numbers n (where $n > 0$) a description of this sequence.]

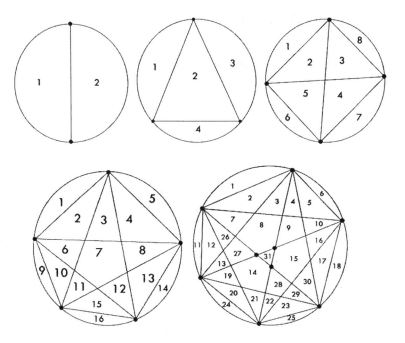

Figure 3-4

To summarize this circle partitioning we will make a chart (figure 3-5) of the number of regions into which a circle can be partitioned by joining the highlighted points on the circle. If you do this independently for further cases, make sure no three lines are concurrent (i.e., meet at one point), or else you will lose a region.

Number of points on the circle	Number of regions into which the circle is partitioned
1	1
2	2
3	4
4	8
5	16
6	31
7	57
8	99

Figure 3-5

Notice that the number of regions listed on the chart gives us (what originally appeared to be) this unusual sequence.

Fibonacci Differences

By now you might be asking how this relates to the Fibonacci numbers. First of all, we know that the Fibonacci sequence does not have a common difference between its terms. After our previous discussion on sequences, this might "delegitimatize" the Fibonacci sequence in your mind. But, before drawing such rash conclusions, let's take a look at the differences between the terms of the Fibonacci sequence (figure 3-6).

Original Sequence	1		1		2		3		5		8		13		21		34		55		89		144
First Differences		0		1		1		2		3		5		8		13		21		34		55	
Second Differences				0		1		1		2		3		5		8		13		21		34
Third Differences						0		1		1		2		3		5		8		13		
Fourth Differences								0		1		1		2		3		5		8		

Figure 3-6

Using the same procedure for inspecting the above sequences, we quickly notice that everywhere on this chart of differences, you will see the Fibonacci sequence. Its appearance could overwhelm us. It is in the first, second, third, fourth, . . . differences, as well as along the diagonals. It appears to reproduce itself in almost all directions! Such a characteristic is one of the many reasons that this sequence has these unusual properties.

If we complete the chart in figure 3-6 using the relationship $F_n = F_{n+2} - F_{n+1}$, which comes from the definition of the Fibonacci numbers (i.e., $F_n + F_{n+1} = F_{n+2}$), the list of differences would be:

Fibonacci sequence: 1, 1, 2, 3, 5, 8, 13, . . .
sequence of first differences: 0, 1, 1, 2, 3, 5, 8, 13, . . .
sequence of second differences: 1, 0, 1, 1, 2, 3, 5, 8, 13, . . .
sequence of third differences: −1, 1, 0, 1, 1, 2, 3, 5, 8, 13, . .
sequence of fourth differences: 2, −1, 1, 0, 1, 1, 2, 3, 5, 8, 13, . . .

(Notice that $F_0 = 0$, $F_{-1} = 1$, $F_{-2} = -1$, . . .)

Furthermore, the sequences of sums generate the Fibonacci sequence again, although a bit shifted over (see figure 3-7).

Sequence of Sums

Label	C1	C2	C3	C4	C5	C6	C7	C8	C9	C10	C11
											10,946
										4,181	6,765
									1,597	2,584	4,181
								610	987	1,597	2,584
							233	377	610	987	1,597
						89	144	233	377	610	987
					34	55	89	144	233	377	610
				13	21	34	55	89	144	233	377
$a + b$			5	8	13	21	34	55	89	144	233
a b		2	3	5	8	13	21	34	55	89	144
F_n	1	1	2	3	5	8	13	21	34	55	89
a b		0	1	1	2	3	5	8	13	21	34
$b - a$			1	0	1	1	2	3	5	8	13
				−1	1	0	1	1	2	3	5
					2	−1	1	0	1	1	2
						−3	2	−1	1	0	1
							5	−3	2	−1	1
								−8	5	−3	2
									13	−8	5
										−21	13
											34

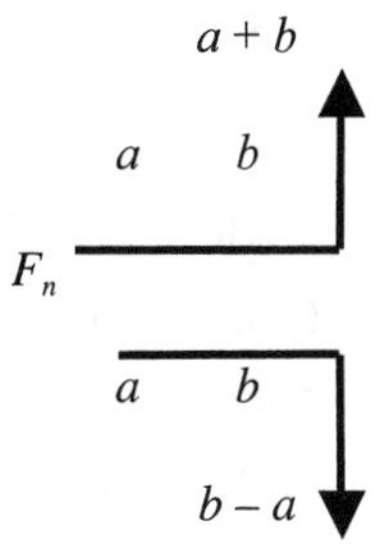

Sequence of Differences

Figure 3-7

Their appearance in the realm of sequence differences is quite amazing, but how does the Fibonacci sequence relate to the other sequences we just inspected? Or put another way: does the Fibonacci sequence relate to these seemingly unrelated sequences? If the past gives us a clue about the future, the answer should be an unexpected "yes." Let's see if we can draw a connection between these three unrelated sequences.

The sequence of powers of 2:

$$1, 2, 4, 8, 16, 32, 64, 128, 256, 512, 1{,}024, \ldots$$

The sequence of circle partitions:

$$1, 2, 4, 8, 16, 31, 57, 99, 163, 256, 386, \ldots$$

The Fibonacci sequence:

$$1, 1, 2, 3, 5, 8, 13, 21, 34, 55, 89, 144, 233, \ldots$$

To do this we will introduce you to the Pascal triangle, named after the famous French mathematician Blaise Pascal (1623–1662), who developed his "arithmetical triangle" as an aid for his work on probability. Pascal showed his genius for mathematics as a child. Even though mathematics books were forbidden to him as a child, his desire and talent for mathematics could not dampen. Through his own inventive construction, he independently proved Euclid's thirty-second proposition, establishing the angle sum of a plane triangle. In order to help his father process huge amounts of data in his position as tax commissioner, he developed a mechanical calculator (called "Pascaline") that by 1645 was on the market for purchase.

The Pascal Triangle

In 1653 Pascal invented his arithmetical triangle,[2] although it was not published until after his death. This famous triangular ar-

---

2. This triangular arrangement of numbers was first described by the Arabian mathematician Omar Khayyam (1048–1122), but first appeared in print in a 1303 manuscript "The Valuable Mirror of the Four Elements" by the Chinese mathematician Chu Shih-Chieh (1270–1330). Yet in the Western world Pascal is credited with having discovered it without reference to any previous documents.

rangement of numbers (figure 3-8) is formed by beginning at the top with 1, then the second row has 1, 1, then the third row is obtained by placing 1s at the end positions and adding the two numbers in the second row (1 + 1 = 2) to get the 2, placed below the interval between the 1s of the previous row. The fourth row is obtained the same way: after the end 1s are placed, the 3s are gotten from the sum of the two numbers above (to the right and left); that is, 1 + 2 = 3 and 2 + 1 = 3. This pattern then continues to get the succeeding rows.

The Pascal Triangle

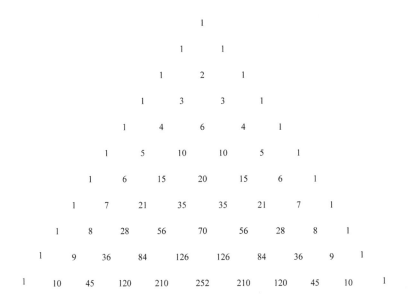

Figure 3-8

Let's look at some of the hidden gems in the Pascal triangle (figure 3-8). If we take the sum of the numbers in each row, we notice that we have arrived at the successive powers of 2 (figure 3-9). Namely,

$$1 = 1 = 2^0,$$

$$1 + 1 = 2 = 2^1,$$

$$1 + 2 + 1 = 4 = 2^2,$$

$$1 + 3 + 3 + 1 = 8 = 2^3,$$

$$1 + 4 + 6 + 4 + 1 = 16 = 2^4,$$

$$1 + 5 + 10 + 10 + 5 + 1 = 32 = 2^5,$$

$$1 + 6 + 15 + 20 + 15 + 6 + 1 = 64 = 2^6,$$

$$1 + 7 + 21 + 35 + 35 + 21 + 7 + 1 = 128 = 2^7,$$

$$1 + 8 + 28 + 56 + 70 + 56 + 28 + 8 + 1 = 256 = 2^8,$$

$$1 + 9 + 36 + 84 + 126 + 126 + 84 + 36 + 9 + 1 = 512 = 2^9,$$

$$1 + 10 + 45 + 120 + 210 + 252 + 210 + 120 + 45 + 10 + 1 = 1{,}024 = 2^{10}$$

This is the first sequence of the three we will connect through the Pascal triangle.

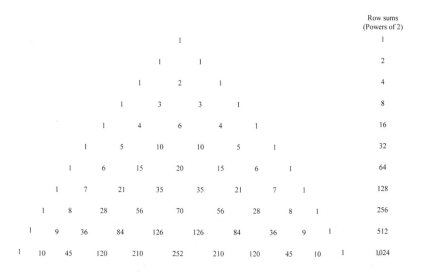

Figure 3-9

Now, consider only the horizontal sums of the rows in figure 3-10 that are to the right of the bold line drawn through the Pascal triangle. There, strangely enough, the sequence of circle partitions emerges!

$$1, 2, 4, 8, 16, 31, 57, 99, 163$$

What a coincidence! So the Pascal triangle provides us with a connection between the first two of the sequences that we wanted to connect.

The Pascal triangle contains many unexpected number relationships beyond the characteristics that Pascal intended in his original use of the triangle. The ubiquity of this triangular arrangement of numbers merits just a bit of further elaboration before we relate the third of our sequences of numbers—the Fibonacci numbers—to the others. Pascal's original use was to exhibit the coefficients of successive terms of a binomial expansion.

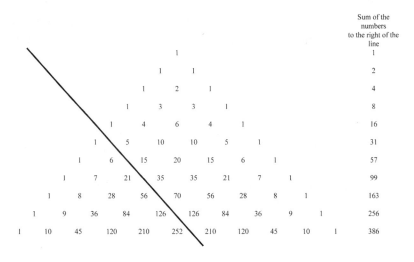

Figure 3-10

That is, taking a binomial such as $(a + b)$ to successively higher powers.

$(a + b)^0 = 1$

$(a + b)^1 = a + b$

$(a + b)^2 = a^2 + 2ab + b^2$

$(a + b)^3 = a^3 + 3a^2b + 3ab^2 + b^3$

$(a + b)^4 = a^4 + 4a^3b + 6a^2b^2 + 4ab^3 + b^4$

$(a + b)^5 = a^5 + 5a^4b + 10a^3b^2 + 10a^2b^3 + 5ab^4 + b^5$

$(a + b)^6 = a^6 + 6a^5b + 15a^4b^2 + 20a^3b^3 + 15a^2b^4 + 6ab^5 + b^6$

$(a + b)^7 = a^7 + 7a^6b + 21a^5b^2 + 35a^4b^3 + 35a^3b^4 + 21a^2b^5 + 7ab^6 + b^7$

$(a + b)^8 = a^8 + 8a^7b + 28a^6b^2 + 56a^5b^3 + 70a^4b^4 + 56a^3b^5 + 28a^2b^6 + 8ab^7 + b^8$

$(a + b)^9 = a^9 + 9a^8b + 36a^7b^2 + 84a^6b^3 + 126a^5b^4 + 126a^4b^5 + 84a^3b^6 + 36a^2b^7$
$\qquad + 9ab^8 + b^9$

$(a + b)^{10} = a^{10} + 10a^9b + 45a^8b^2 + 120a^7b^3 + 210a^6b^4 + 252a^5b^5 + 210a^4b^6 + 120a^3b^7$
$\qquad + 45a^2b^8 + 10ab^9 + b^{10}$

$\ldots$

Notice how the coefficients of each binomial-expansion line is also represented as a row of the Pascal triangle. This allows us to expand a binomial without actually multiplying it by itself many times to get the end result. There is a pattern also among the variables' exponents: one descends while the other ascends in value—each time keeping the sum of the exponents constant—making it equal to the exponent of the original binomial taken to a power.[3]

---

3. Some may be interested in the general term of the expansion of $(a + b)$, so we offer it here:

$$\left(a+b\right)^n = \binom{n}{0}a^n + \binom{n}{1}a^{n-1}b + \binom{n}{2}a^{n-2}b^2 + \ldots + \binom{n}{n-2}a^2b^{n-2} + \binom{n}{n-1}ab^{n-1} + \binom{n}{n}b^n,$$

where $\binom{n}{k} = \dfrac{n!}{k!\cdot(n-k)!}$, and $n! = 1 \cdot 2 \cdot 3 \cdot 4 \ldots n$.

We should note that we define $\binom{n}{0} = 1$.

For example, to get the fourth entry in the eighth row, we use

$$\binom{n}{k} = \binom{7}{3} = \frac{7!}{3!\cdot(7-3)!} = \frac{7!}{3!\cdot4!} = \frac{4!\cdot5\cdot6\cdot7}{3!\cdot4!} = \frac{5\cdot6\cdot7}{3!} = \frac{5\cdot6\cdot7}{1\cdot2\cdot3} = 35.$$

By the way, the binomial coefficient $\binom{n}{k}$ tells us how many ways n coin tosses come up with k heads.

There are many more intriguing surprises embedded in this wonderful triangular arrangement of numbers. Consider the Pascal triangle in figure 3-11 with the various cells shaded.

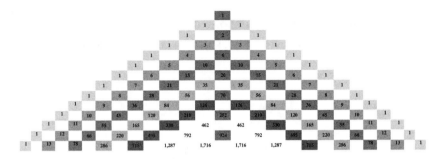

Figure 3-11

First we notice the list of 1s on both sides of the triangle:

1, 1, 1, 1, 1, 1, 1, 1, 1, 1, 1, 1, . . .

The next list in the parallel direction is the list of natural numbers:

1, 2, 3, 4, 5, 6, 7, 8, 9, 10, 11, 12, 13, . . .

Continuing in this pattern (i.e., parallel lists) we get the list of triangular numbers:[4]

1, 3, 6, 10, 15, 21, 28, 36, 45, 55, 66, 78, . . .

The next list of numbers in the same direction is the list of tetrahedral numbers:[5]

1, 4, 10, 20, 35, 56, 84, 120, 165, 220, 286, . . .

We then come to a lesser-known sequence of numbers that exists in the fourth dimension called pentatop numbers:

1, 5, 15, 35, 70, 126, 210, 330, 495, 715, . . .

Suppose you would like to add the numbers in one of these sequences of numbers. All you need to do is locate the number in the cell that is immediately below and to the right (or left) of the last

---

4. Triangular numbers represent the number of points that can be arranged to form an equilateral triangle.
5. Tetrahedral numbers represent the number of points that can be arranged to form a regular tetrahedron.

number in the list of numbers you are adding. For example, consider the triangular numbers (see figure 3-12):

$$1 + 3 + 6 + 10 + 15 + 21 = 56$$

The same holds true for the sum of tetrahedral numbers:

$$1 + 3 + 6 + 10 + 15 + 21 + 28 + 36 + 45 + 55 + 66 = 286$$

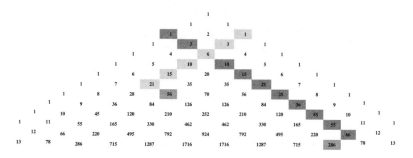

Figure 3-12

Fibonacci Numbers and the Pascal Triangle

By now you must be wondering where on the Pascal triangle are the Fibonacci numbers hiding, for, after all, isn't that what you expect by now? They seem to crop up everywhere, even when we least expect to see them. Yes, indeed, the Fibonacci numbers are embedded on the Pascal triangle. Look at the sums of the numbers along each of the indicated lines (see figure 3-13). There you have the Fibonacci numbers! We have then connected, with the help of the Pascal triangle, the three sequences we sought to relate.

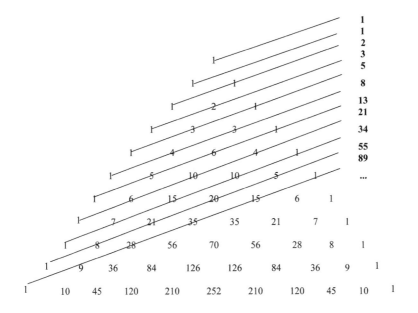

Figure 3-13

This may be easier to visualize if we left justify the triangular arrangement of numbers as in figure 3-14:

							Sum
1							1
1							1
1	1						2
1	2						3
1	3	1					5
1	4	3					8
1	5	6	1				13
1	6	10	4				21
1	7	15	10	1			34
1	8	21	20	5			55
1	9	28	35	15	1		89
1	10	36	56	35	6		144
1	11	45	84	70	21	1	233

. . .

Figure 3-14

Ron Knott, a British mathematician who has provided some interesting insights into the Fibonacci numbers, found the Fibonacci numbers appearing as sums of "rows" in Pascal's triangle.

	0	1	2	3	4	5	6	7	8	9	10	11	12	13	14	15	16	17	18	19	20
0	1																				
1		1	1																		
2			1	2	1																
3				1	3	3	1														
4					1	4	6	4	1												
5						1	5	10	10	5	1										
6							1	6	15	20	15	6	1								
7								1	7	21	35	35	21	7	1						
8									1	8	28	56	70	56	28	8	1				
9										1	9	36	84	126	126	84	36	9	1		
10											1	10	45	120	210	252	210	120	45	10	1
11												1	11	55	165	330	462	462	330	165	55
12													1	12	66	220	495	792	924	792	495
13														1	13	78	286	715	1287	1716	1716
14															1	14	91	364	1001	2002	3003
15																1	15	105	455	1365	3003
16																	1	16	120	560	1820
17																		1	17	136	680
18																			1	18	153
19																				1	19
20																					1
	1	1	2	3	5	8	13	21	34	55	89	144	233	377	610	987	1597	2584	4181	6765	10946

Figure 3-15

By drawing Pascal's triangle with all the rows moved over by one place, we have a clearer arrangement, which shows the Fibonacci numbers as sums of columns. (See the last row in figure 3-15.)

We could also rewrite this version of the Pascal triangle—moving each row over to the right by one column—and writing each row twice. By doing this we find that the column sums again yield the Fibonacci numbers (see figure 3-16).

	2	3	4	5	6	7	8	9	10	11	12	13	14	15	16	17	18	19	20	21	22
0	1																				
		1																			
1		1	1																		
			1	1																	
2			1	2	1																
				1	2	1															
3				1	3	3	1														
					1	3	3	1													
4					1	4	6	4	1												
						1	4	6	4	1											
5						1	5	10	10	5	1										
							1	5	10	10	5	1									
6							1	6	15	20	15	6	1								
								1	6	15	20	15	6	1							
7								1	7	21	35	35	21	7	1						
									1	7	21	35	35	21	7	1					
8									1	8	28	56	70	56	28	8	1				
										1	8	28	56	70	56	28	8	1			
9										1	9	36	84	126	126	84	36	9	1		
											1	9	36	84	126	126	84	36	9	1	
10											1	10	45	120	210	252	210	120	45	10	1
												1	10	45	120	210	252	210	120	45	10
11												1	11	55	165	330	462	462	330	165	55
													1	11	55	165	330	462	462	330	165
12													1	12	66	220	495	792	924	792	495
														1	12	66	220	495	792	924	792
13														1	13	78	286	715	1287	1716	1716
															1	13	78	286	715	1287	1716
14															1	14	91	364	1001	2002	3003
																1	14	91	364	1001	2002
15																1	15	105	455	1365	3003
																	1	15	105	455	1365
16																	1	16	120	560	1820
																		1	16	120	560
17																		1	17	136	680
																			1	17	136
18																			1	18	153
																				1	18
19																				1	19
																					1
20																					1
	1	2	3	5	8	13	21	34	55	89	144	233	377	610	987	1597	2584	4181	6765	10946	17711

Figure 3-16

If we revert once again to the left-justified Pascal triangle (figure 3-17), we can easily discover a nice group of palindromic numbers:[6] 1,001, 2,002, 3,003, 5,005, and 8,008.

---

6. Palindromic numbers are those that read the same in both directions, such as 3,003, or from the Pascal triangle: 1,331 and 14,641.

```
1
1  1
1  2   1
1  3   3    1
1  4   6    4    1
1  5   10   10   5    1
1  6   15   20   15   6    1
1  7   21   35   35   21   7    1
1  8   28   56   70   56   28   8    1
1  9   36   84   126  126  84   36   9    1
1 10   45   120  210  252  210  120  45   10   1
1 11   55   165  330  462  462  330  165  55   11   1
1 12   66   220  495  792  924  792  495  220  66   12   1
1 13   78   286  715  1287 1716 1716 1287 715  286  78   13   1
1 14   91   364  1001 2002 3003 3432 3003 2002 1001 364  91   14   1
1 15   105  455  1365 3003 5005 6435 6435 5005 3003 1365 455  105  15   1
1 16   120  560  1820 4368 8008 11440 12870 11440 8008 4368 1820 560  120  16   1
1 17   136  680  2380 6188 12376 19448 24310 24310 19448 12376 6188 2380 680  136  17   1
1 18   153  816  3060 8568 18564 31824 43758 48620 43758 31824 18564 8568 3060 816  153  18   1
1 19   171  969  3876 11628 27132 50388 75582 92378 92378 75582 50388 27132 11628 3876 969  171  19   1
1 20   190  1140 4845 15504 38760 77520 125970 167960 184756 167960 125970 77520 38760 15504 4845 1140 190  20   1
```

Figure 3-17

In 1971, David Singmaster—an American mathematician living in London, who has analyzed many aspects of recreational mathematics to secure them into the realm of serious mathematics[7]—discovered[8] that the fifteenth row is the only one in which three consecutive numbers will be in the ratio 1 : 2 : 3. They are: 1,001, 2,002, and 3,003. It should not be lost on the reader that coincidentally the numbers under these three are also palindromes in a Fibonacci ratio! (See figure 3-18.)

Notice that $1,001 = 1 \cdot 1,001$; $2,002 = 2 \cdot 1,001$; $3,003 = 3 \cdot 1,001$; $5,005 = 5 \cdot 1,001$; and $8,008 = 8 \cdot 1,001$.

They all have 1,001 as their common factor.

```
1,001              2,002              3,003
          3,003              5,005
                   8,008
```

Figure 3-18

---

7. Such as the Rubic's Cube.
8. *American Mathematical Monthly* 78 (1971): 385–86.

Lest you worry about our quest for a relationship between the Fibonacci numbers and the Pascal triangle, consider this. Using our earlier symbols to represent the Fibonacci numbers, and considering that we can extend the Fibonacci number sequence in the negative direction as follows:[9]

$$\ldots, 13, -8, 5, -3, 2, -1, 1, 0, 1, 1, 2, 3, 5, 8, 13, \ldots,$$

we can also represent the Fibonacci numbers in terms of the members of the Pascal triangle as shown in figure 3-19.

$$F_n = 1 \cdot F_n$$

$$F_{n+1} = 1 \cdot F_n + 1 \cdot F_{n-1}$$

$$F_{n+2} = 1 \cdot F_n + 2 \cdot F_{n-1} + 1 \cdot F_{n-2}$$

$$F_{n+3} = 1 \cdot F_n + 3 \cdot F_{n-1} + 3 \cdot F_{n-2} + 1 \cdot F_{n-3}$$

$$F_{n+4} = 1 \cdot F_n + 4 \cdot F_{n-1} + 6 \cdot F_{n-2} + 4 \cdot F_{n-3} + 1 \cdot F_{n-4}$$

$$F_{n+5} = 1 \cdot F_n + 5 \cdot F_{n-1} + 10 \cdot F_{n-2} + 10 \cdot F_{n-3} + 5 \cdot F_{n-4} + 1 \cdot F_{n-5}$$

$$F_{n+6} = 1 \cdot F_n + 6 \cdot F_{n-1} + 15 \cdot F_{n-2} + 20 \cdot F_{n-3} + 15 \cdot F_{n-4} + 6 \cdot F_{n-5} + 1 \cdot F_{n-6}$$

$$F_{n+7} = 1 \cdot F_n + 7 \cdot F_{n-1} + 21 \cdot F_{n-2} + 35 \cdot F_{n-3} + 35 \cdot F_{n-4} + 21 \cdot F_{n-5} + 7 \cdot F_{n-6} + 1 \cdot F_{n-7}$$

$$F_{n+8} = 1 \cdot F_n + 8 \cdot F_{n-1} + 28 \cdot F_{n-2} + 56 \cdot F_{n-3} + 70 \cdot F_{n-4} + 56 \cdot F_{n-5} + 28 \cdot F_{n-6} + 8 \cdot F_{n-7} + 1 \cdot F_{n-8}$$

. . .

Figure 3-19

At first glance, this may look artificial and contrived, but reserve your judgment until you begin to substitute the Fibonacci values to verify the relationship.

---

9. See, for instance, the sequence of fourth differences and the diagonal in figure 3-7 (from left to right).

You can see from the negatively extended Fibonacci sequence that $F_0 = 0$, $F_{-2n} = -F_{2n}$, and $F_{-2n+1} = F_{2n-1}$ for all natural numbers n. Furthermore, the recursive definition from before (the actual basis for the Fibonacci numbers), $F_{k+2} = F_{k+1} + F_k$, still holds true for all integer values of k.

To get a better understanding about this unusual relationship, let's check the case where $n = 5$ (figure 3-20):

$$F_5 = 1 \cdot F_5$$

$$F_6 = 1 \cdot F_5 + 1 \cdot F_4$$

$$F_7 = 1 \cdot F_5 + 2 \cdot F_4 + 1 \cdot F_3$$

$$F_8 = 1 \cdot F_5 + 3 \cdot F_4 + 3 \cdot F_3 + 1 \cdot F_2$$

$$F_9 = 1 \cdot F_5 + 4 \cdot F_4 + 6 \cdot F_3 + 4 \cdot F_2 + 1 \cdot F_1$$

$$F_{10} = 1 \cdot F_5 + 5 \cdot F_4 + 10 \cdot F_3 + 10 \cdot F_2 + 5 \cdot F_1 + 1 \cdot F_0$$

$$F_{11} = 1 \cdot F_5 + 6 \cdot F_4 + 15 \cdot F_3 + 20 \cdot F_2 + 15 \cdot F_1 + 6 \cdot F_0 + 1 \cdot F_{-1}$$

$$F_{12} = 1 \cdot F_5 + 7 \cdot F_4 + 21 \cdot F_3 + 35 \cdot F_2 + 35 \cdot F_1 + 21 \cdot F_0 + 7 \cdot F_{-1} + 1 \cdot F_{-2}$$

$$F_{13} = 1 \cdot F_5 + 8 \cdot F_4 + 28 \cdot F_3 + 56 \cdot F_2 + 70 \cdot F_1 + 56 \cdot F_0 + 28 \cdot F_{-1} + 8 \cdot F_{-2} + 1 \cdot F_{-3}$$

. . .

Figure 3-20

To get a better "feel" for this, we will substitute the appropriate Fibonacci numbers in the listing shown in figure 3-20. This allows us to calculate the "generated" Fibonacci numbers at the right in figure 3-21.

$$F_5 = 1 \cdot 5 \qquad\qquad\qquad\qquad\qquad\qquad = 5$$

$$F_6 = 1 \cdot 5 + 1 \cdot 3 \qquad\qquad\qquad\qquad\qquad = 8$$

$$F_7 = 1 \cdot 5 + 2 \cdot 3 + 1 \cdot 2 \qquad\qquad\qquad\qquad = 13$$

$$F_8 = 1 \cdot 5 + 3 \cdot 3 + 3 \cdot 2 + 1 \cdot 1 \qquad\qquad\qquad = 21$$

$$F_9 = 1 \cdot 5 + 4 \cdot 3 + 6 \cdot 2 + 4 \cdot 1 + 1 \cdot 1 \qquad\qquad = 34$$

$$F_{10} = 1 \cdot 5 + 5 \cdot 3 + 10 \cdot 2 + 10 \cdot 1 + 5 \cdot 1 + 1 \cdot 0 \qquad = 55$$

$$F_{11} = 1 \cdot 5 + 6 \cdot 3 + 15 \cdot 2 + 20 \cdot 1 + 15 \cdot 1 + 6 \cdot 0 + 1 \cdot 1 \qquad = 89$$

$$F_{12} = 1 \cdot 5 + 7 \cdot 3 + 21 \cdot 2 + 35 \cdot 1 + 35 \cdot 1 + 21 \cdot 0 + 7 \cdot 1 + 1 \cdot (-1) \quad = 144$$

$$F_{13} = 1 \cdot 5 + 8 \cdot 3 + 28 \cdot 2 + 56 \cdot 1 + 70 \cdot 1 + 56 \cdot 0 + 28 \cdot 1 + 8 \cdot (-1) + 1 \cdot 2 = 233$$

. . .

Figure 3-21

Lucas Numbers and the Pascal Triangle

As if this weren't enough, the Pascal triangle also enables us to connect in yet another way the Fibonacci numbers to the Lucas numbers
that we mentioned in chapter 1. Recall that Edouard Lucas (1842–
1891), a French mathematician, developed the sequence of numbers
with the same recursive rule as the Fibonacci numbers, but he began
with 1 and 3 instead of 1 and 1. So he got the sequence 1, 3, 4, 7, 11,
18, . . . instead of the Fibonacci numbers 1, 1, 2, 3, 5, 8, ... In figure 3-22 we list the Lucas numbers. To make things even more interesting, however, we will begin the sequence one number earlier
than the traditional beginning. That is, instead of beginning with the
first Lucas number, L_1, we will begin with L_0, which is 2.

We will represent the nth Lucas number as L_n so that $L_1 = 1$,
$L_2 = 3$ (and as we said earlier, $L_0 = 2$), and $L_{n+2} = L_{n+1} + L_n$.

As we begin to relate the Fibonacci and the Lucas sequences, we
should realize a direct connection: The nth Lucas number ($n \geq 0$) is

equal to the sum of the $(n-1)$st Fibonacci number and the $(n+1)$st Fibonacci number. Symbolically, that is written as: $L_n = F_{n-1} + F_{n+1}$.

n	L_n
0	2
1	1
2	3
3	4
4	7
5	11
6	18
7	29
8	47
9	76
10	123
11	199
12	322
13	521
14	843
15	1,364
16	2,207
17	3,571
18	5,778
19	9,349
20	15,127

Figure 3-22

In figure 3-23 you can observe this relationship ($L_n = F_{n-1} + F_{n+1}$) with the two sequences placed next to each other.

n	0	1	2	3	4	5	6	7	8	9	10	11	12	13	14	15	16	17	18	19	20
F_{n+1}		1	2	3	5	8	13	21	34	55	89	144	233	377	610	987	1,597	2,584	4,181	6,765	10,946
+																					
F_{n-1}			1	1	2	3	5	8	13	21	34	55	89	144	233	377	610	987	1,597	2,584	4,181
=																					
L_n		1	3	4	7	11	18	29	47	76	123	199	322	521	843	1,364	2,207	3,571	5,778	9,349	15,127

Figure 3-23

Since these two sequences are so closely related, it is only fitting, and not particularly surprising, that we can also represent the Lucas numbers in the arrangement of the Pascal triangle—just as we did with the Fibonacci numbers (see figure 3-24). Keep in mind that $L_0 = 2$, $L_{-2n} = L_{2n}$, and $L_{-2n+1} = -L_{2n-1}$, for all natural numbers n.

$$L_n = 1 \cdot L_n$$

$$L_{n+1} = 1 \cdot L_n + 1 \cdot L_{n-1}$$

$$L_{n+2} = 1 \cdot L_n + 2 \cdot L_{n-1} + 1 \cdot L_{n-2}$$

$$L_{n+3} = 1 \cdot L_n + 3 \cdot L_{n-1} + 3 \cdot L_{n-2} + 1 \cdot L_{n-3}$$

$$L_{n+4} = 1 \cdot L_n + 4 \cdot L_{n-1} + 6 \cdot L_{n-2} + 4 \cdot L_{n-3} + 1 \cdot L_{n-4}$$

$$L_{n+5} = 1 \cdot L_n + 5 \cdot L_{n-1} + 10 \cdot L_{n-2} + 10 \cdot L_{n-3} + 5 \cdot L_{n-4} + 1 \cdot L_{n-5}$$

$$L_{n+6} = 1 \cdot L_n + 6 \cdot L_{n-1} + 15 \cdot L_{n-2} + 20 \cdot L_{n-3} + 15 \cdot L_{n-4} + 6 \cdot L_{n-5} + 1 \cdot L_{n-6}$$

$$L_{n+7} = 1 \cdot L_n + 7 \cdot L_{n-1} + 21 \cdot L_{n-2} + 35 \cdot L_{n-3} + 35 \cdot L_{n-4} + 21 \cdot L_{n-5} + 7 \cdot L_{n-6} + 1 \cdot L_{n-7}$$

$$L_{n+8} = 1 \cdot L_n + 8 \cdot L_{n-1} + 28 \cdot L_{n-2} + 56 \cdot L_{n-3} + 70 \cdot L_{n-4} + 56 \cdot L_{n-5} + 28 \cdot L_{n-6} + 8 \cdot L_{n-7} + 1 \cdot L_{n-8}$$

. . .

Figure 3-24

Moreover, the recursive definition, $L_{n+2} = L_{n+1} + L_n$, which is the basis for the Lucas and the Fibonacci numbers, still holds true for all integer values of n. We can see that in figure 3-25.

We are now ready to experience the beautiful Pascal triangle relationship with actual Lucas numbers. For that we consider the case where $n = 5$. To accomplish this, we will substitute the actual Lucas numbers in the listing of figure 3-24, so that we get the listing shown in figure 3-26 with the Lucas numbers generated at the right.

n	L_n
0	2
−1	−1
−2	3
−3	−4
−4	7
−5	−11
−6	18
−7	−29
−8	47
−9	−76
−10	123
−11	−199
−12	322
−13	−521
−14	843
−15	−1,364
−16	2,207
−17	−3,571

Figure 3-25

$L_5 = 1\cdot11$ $= 11$

$L_6 = 1\cdot11 + 1\cdot7$ $= 18$

$L_7 = 1\cdot11 + 2\cdot7 + 1\cdot4$ $= 29$

$L_8 = 1\cdot11 + 3\cdot7 + 3\cdot4 + 1\cdot3$ $= 47$

$L_9 = 1\cdot11 + 4\cdot7 + 6\cdot4 + 4\cdot3 + 1\cdot1$ $= 76$

$L_{10} = 1\cdot11 + 5\cdot7 + 10\cdot4 + 10\cdot3 + 5\cdot1 + 1\cdot2$ $= 123$

$L_{11} = 1\cdot11 + 6\cdot7 + 15\cdot4 + 20\cdot3 + 15\cdot1 + 6\cdot2 + 1\cdot(-1)$ $= 199$

$L_{12} = 1\cdot11 + 7\cdot7 + 21\cdot4 + 35\cdot3 + 35\cdot1 + 21\cdot2 + 7\cdot(-1) + 1\cdot3$ $= 322$

$L_{13} = 1\cdot11 + 8\cdot7 + 28\cdot4 + 56\cdot3 + 70\cdot1 + 56\cdot2 + 28\cdot(-1) + 8\cdot3 + 1\cdot(-4) = 521$

...

Figure 3-26

We can also get the initial Lucas numbers in the same way. Consider the case where $n = 1$, and substitute the values of the appropriate Lucas numbers in figure 3-24. We then get figure 3-27, with the Lucas numbers generated at the right.

$L_1 = \mathbf{1 \cdot 1}$ $= 1$

$L_2 = \mathbf{1 \cdot 1} + \mathbf{1 \cdot 2}$ $= 3$

$L_3 = \mathbf{1 \cdot 1} + \mathbf{2 \cdot 2} + \mathbf{1 \cdot}(-1)$ $= 4$

$L_4 = \mathbf{1 \cdot 1} + \mathbf{3 \cdot 2} + \mathbf{3 \cdot}(-1) + \mathbf{1 \cdot 3}$ $= 7$

$L_5 = \mathbf{1 \cdot 1} + \mathbf{4 \cdot 2} + \mathbf{6 \cdot}(-1) + \mathbf{4 \cdot 3} + \mathbf{1 \cdot}(-4)$ $= 11$

$L_6 = \mathbf{1 \cdot 1} + \mathbf{5 \cdot 2} + \mathbf{10 \cdot}(-1) + \mathbf{10 \cdot 3} + \mathbf{5 \cdot}(-4) + \mathbf{1 \cdot 7}$ $= 18$

$L_7 = \mathbf{1 \cdot 1} + \mathbf{6 \cdot 2} + \mathbf{15 \cdot}(-1) + \mathbf{20 \cdot 3} + \mathbf{15 \cdot}(-4) + \mathbf{6 \cdot 7} + \mathbf{1 \cdot}(-11)$ $= 29$

$L_8 = \mathbf{1 \cdot 1} + \mathbf{7 \cdot 2} + \mathbf{21 \cdot}(-1) + \mathbf{35 \cdot 3} + \mathbf{35 \cdot}(-4) + \mathbf{21 \cdot 7} + \mathbf{7 \cdot}(-11) + \mathbf{1 \cdot 18}$ $= 47$

$L_9 = \mathbf{1 \cdot 1} + \mathbf{8 \cdot 2} + \mathbf{28 \cdot}(-1) + \mathbf{56 \cdot 3} + \mathbf{70 \cdot}(-4) + \mathbf{56 \cdot 7} + \mathbf{28 \cdot}(-11) + \mathbf{8 \cdot 18} + \mathbf{1 \cdot}(-29) = 76$

...

Figure 3-27

For the sake of consistency, we would want to be able to find the Lucas numbers on the Pascal triangle. To do so, we will modify the original Pascal triangle by replacing the right side 1s with 2s and then carrying on the calculation as we did for the original Pascal triangle to generate the rows. See figure 3-28.

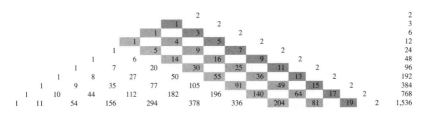

Figure 3-28

Before we search for the Lucas numbers, we should take note of some interesting features of this type of Pascal triangle. Looking at the diagonal rows in figure 3-28, you can see the odd numbers, the square numbers, and the pyramidal numbers. In addition, if you inspect the diagonal rows in the other direction, you will find a progression followed by sequences of progressive differences, a natural result of the Pascal triangle's construction.

As we found the Fibonacci numbers in the original Pascal triangle, so, too, will we now locate the Lucas numbers in the second type of Pascal triangle. To locate the Lucas numbers, notice the sum of the numbers along each of the lines drawn in figure 3-29. This is analogous to the way in which we located the Fibonacci numbers on the Pascal triangle.

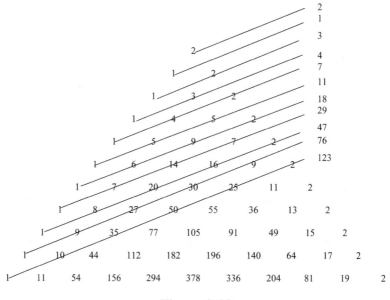

Figure 3-29

By writing this triangular arrangement of numbers in a left-justified fashion (figure 3-30), we can get the column sums to generate the Lucas numbers.

	0	1	2	3	4	5	6	7	8	9	10	11	12	13	14	15	16	17	18	19	20
0	2																				
1		1	2																		
2			1	3	2																
3				1	4	5	2														
4					1	5	9	7	2												
5						1	6	14	16	9	2										
6							1	7	20	30	25	11	2								
7								1	8	27	50	55	36	13	2						
8									1	9	35	77	105	91	49	15	2				
9										1	10	44	112	182	196	140	64	17	2		
10											1	11	54	156	294	378	336	204	81	19	2
11												1	12	65	210	450	672	714	540	285	100
12													1	13	77	275	660	1122	1386	1254	825
13														1	14	90	352	935	1782	2508	2640
14															1	15	104	442	1287	2717	4290
15																1	16	119	546	1729	4004
16																	1	17	135	665	2275
17																		1	18	152	800
18																			1	19	170
19																				1	20
20																					1
	2	1	3	4	7	11	18	29	47	76	123	199	322	521	843	1364	2207	3571	5778	9349	15127

Figure 3-30

If we repeat this with double entries, as we have for the Fibonacci numbers on the Pascal triangle, we can again generate the Lucas numbers by taking the column sums, as in figure 3-31.

	0	1	2	3	4	5	6	7	8	9	10	11	12	13	14	15	16	17	18	19	20
0	2																				
		2																			
1		1	2																		
			1	2																	
2			1	3	2																
				1	3	2															
3				1	4	5	2														
					1	4	5	2													
4					1	5	9	7	2												
						1	5	9	7	2											
5						1	6	14	16	9	2										
							1	6	14	16	9	2									
6							1	7	20	30	25	11	2								
								1	7	20	30	25	11	2							
7								1	8	27	50	55	36	13	2						
									1	8	27	50	55	36	13	2					
8									1	9	35	77	105	91	49	15	2				
										1	9	35	77	105	91	49	15	2			
9										1	10	44	112	182	196	140	64	17	2		
											1	10	44	112	182	196	140	64	17	2	
10											1	11	54	156	294	378	336	204	81	19	2
												1	11	54	156	294	378	336	204	81	19
11												1	12	65	210	450	672	714	540	285	100
													1	12	65	210	450	672	714	540	285
12													1	13	77	275	660	1122	1386	1254	825
														1	13	77	275	660	1122	1386	1254
13														1	14	90	352	935	1782	2508	2640
															1	14	90	352	935	1782	2508
14															1	15	104	442	1287	2717	4290
																1	15	104	442	1287	2717
15																1	16	119	546	1729	4004
																	1	16	119	546	1729
16																	1	17	135	665	2275
																		1	17	135	665
17																		1	18	152	800
																			1	18	152
18																			1	19	170
																				1	19
19																				1	20
																					1
20																					1
	(2)	3	4	7	11	18	29	47	76	123	199	322	521	843	1364	2207	3571	5778	9349	15127	24476

Figure 3-31

Sequences of numbers, as we have just seen, are not always what we might expect upon a first inspection. They can be related even if they appear completely unrelated. We have now taken one of the most ubiquitous arrays of numbers in mathematics, the Pascal triangle, to show how, via this array, number sequences that are seemingly entirely unrelated can in fact have some commonality. The Fibonacci numbers and many other familiar sequences are embedded in the Pascal triangle. The modified Pascal triangle relates the Lucas numbers to other well-known sequences. It shows a very strong connection between the Fibonacci numbers and the Lucas numbers. This is just the beginning. We invite you to search for others in this amazing array of numbers.

Chapter 4

The Fibonacci Numbers and the Golden Ratio

It is time we ought to consider the geometric manifestation of the Fibonacci numbers, the beauty of which has been appreciated in many fields. We will show the relationship between the Fibonacci numbers and the golden ratio, which has generally been observed in the golden section, the golden rectangle, and the golden triangle, as well as other related figures along the way. In any case, this geometry excursion is well worth the journey!

Fibonacci Ratios

In the previous chapter, we viewed the Fibonacci numbers as a sequence, rather than as individual numbers. We noticed how the Fibonacci sequence related to other sequences of numbers that, at the start, seemed completely unrelated to it. We will now inspect the Fibonacci sequence of numbers by considering the relationship of consecutive members of the sequence. Such a relationship is best seen by taking the ratio of these consecutive numbers. We can see that, as the numbers get larger, the ratio seems to approach a specific number. But, as the numbers get larger, what is this number that the ratios are approaching?

$$\frac{F_2}{F_1} = \frac{1}{1} = 1$$

$$\frac{F_3}{F_2} = \frac{2}{1} = 2$$

$$\frac{F_4}{F_3} = \frac{3}{2} = 1.5$$

$$\frac{F_5}{F_4} = \frac{5}{3} = 1.\overline{6}$$

$$\frac{F_6}{F_5} = \frac{8}{5} = 1.6$$

$$\frac{F_7}{F_6} = \frac{13}{8} = 1.625$$

$$\frac{F_8}{F_7} = \frac{21}{13} = 1.\overline{615384}\ [1]$$

$$\frac{F_9}{F_8} = \frac{34}{21} = 1.\overline{619047}$$

$$\frac{F_{10}}{F_9} = \frac{55}{34} = 1.6\overline{1764705882352941}$$

$$\frac{F_{11}}{F_{10}} = \frac{89}{55} = 1.6\overline{18}$$

$$\frac{F_{12}}{F_{11}} = \frac{144}{89} = 1.6\overline{179775280898876404494382022471910112359550}5$$

$$\frac{F_{13}}{F_{12}} = \frac{233}{144} = 1.61\overline{805}$$

$$\frac{F_{14}}{F_{13}} = \frac{377}{233} = 1.618025751072961373390557939914 2 \ldots\ [2]$$

---

1. A reminder: the bar over the digits after the decimal point indicates that these digits repeat endlessly.

2. The period of the decimal expansion of $\frac{377}{233}$ is
6180257510729613733905579399141630901287553648068669527896995708154506437768240343347639484978540772532188841201716738197424892703862660944206008583690987124463519313304721030042918454935622317596566523605150214592274678111587982832, which then repeats endlessly. The period has length 232 [= 233 − 1].

We could have also considered the ratio of the numbers of the Fibonacci sequence in the inverted order. Take a look at the two sets of ratios in figure 4-1. Can you see anything curious about the respective numbers they are approaching? It should begin to become evident with the larger numbers

The Ratios of Consecutive Fibonacci Numbers[3]

$\dfrac{F_{n+1}}{F_n}$	$\dfrac{F_n}{F_{n+1}}$
$\dfrac{1}{1} = 1.000000000$	$\dfrac{1}{1} = 1.000000000$
$\dfrac{2}{1} = 2.000000000$	$\dfrac{1}{2} = 0.500000000$
$\dfrac{3}{2} = 1.500000000$	$\dfrac{2}{3} = 0.666666667$
$\dfrac{5}{3} = 1.666666667$	$\dfrac{3}{5} = 0.600000000$
$\dfrac{8}{5} = 1.600000000$	$\dfrac{5}{8} = 0.625000000$
$\dfrac{13}{8} = 1.625000000$	$\dfrac{8}{13} = 0.615384615$
$\dfrac{21}{13} = 1.615384615$	$\dfrac{13}{21} = 0.619047619$
$\dfrac{34}{21} = 1.619047619$	$\dfrac{21}{34} = 0.617647059$
$\dfrac{55}{34} = 1.617647059$	$\dfrac{34}{55} = 0.618181818$
$\dfrac{89}{55} = 1.618181818$	$\dfrac{55}{89} = 0.617977528$

---

3. Rounded to 9 decimal places.

$$\frac{144}{89} = 1.617977528 \qquad \frac{89}{144} = 0.618055556$$

$$\frac{233}{144} = 1.618055556 \qquad \frac{144}{233} = 0.618025751$$

$$\frac{377}{233} = 1.618025751 \qquad \frac{233}{377} = 0.618037135$$

$$\frac{610}{377} = 1.618037135 \qquad \frac{377}{610} = 0.618032787$$

$$\frac{987}{610} = 1.618032787 \qquad \frac{610}{987} = 0.618034448$$

Figure 4-1

Both columns are each approaching a specific number (with the right column lagging one ratio behind). The left column seems to be approaching the value 1.61803 . . . , while the right column appears to be approaching 0.61803 . . . . As the numbers in the ratio become very large, we might conclude that[4] $\frac{F_{n+1}}{F_n} = \frac{F_{n-1}}{F_n} + 1$. But as the numbers get larger and larger, the one step lag that we observed in figure 4-1 becomes negligible. We can thus say that, in general, $\frac{F_{n+1}}{F_n} \approx \frac{F_n}{F_{n+1}} + 1$. This is called the golden ratio—when the Fibonacci numbers used reach their limitless "largeness."

The Golden Ratio

The limit of this ratio is perhaps one of the most famous numbers in mathematics. Convention has it that the Greek letter ϕ (phi) is used to represent this ratio. There is reason to believe that the letter ϕ was used because it is the first letter of the name of celebrated Greek sculptor Phidias (ca. 490–430 BCE),[5] who produced the

---

4. Even though there is a lag—by one—in the right column, when the numbers get larger this lag becomes relatively insignificant.
5. In Greek: ΦΣΙΔΙΑΣ.

famous statue of Zeus in the Temple of Olympia and supervised the construction of the Parthenon in Athens, Greece. His frequent use of the golden ratio in this glorious building (see chapter 7) is likely the reason for this attribution.

(In a lighthearted way, some like to relate the Fibonacci numbers to ϕ [phi], by saying that they are really called the *Phi*-bonacci numbers or ϕ-bonacci numbers!)

A somewhat more accurate estimate of ϕ would look like this: $\phi \approx 1.61803398874989484482045868343656$, which is what the Fibonacci number ratios in figure 4-1 seem to be getting ever closer to. The unique characteristic that we can observe from this chart is that $\phi = \frac{1}{\phi} + 1$, or $\frac{1}{\phi} = \phi - 1$. This way, we can calculate the value of the reciprocal of ϕ from the value of it given above, by merely subtracting 1, to get: $\frac{1}{\phi} = 0.61803398874989484482045868\ 343656 \ldots$ As a matter of fact, this is the only number where such a relationship is true. This should not be confused with the universally true mathematical relationship that $\phi \cdot \frac{1}{\phi} = 1$. This, of course, is true for all numbers other than zero.

For those wanting a more precise value of ϕ, we offer it here to one thousand places:

$\phi = 1.6180339887498948482045868343656381177203091798057628621$
354486227052604628189024497072072041893911374847540880753868917521266338622235369317931800607667263544333890865959395829056383226613199282902678806752087668925017116962070322210432162695486262963136144381497587012203408058879544547492461856953648644492410443207713449470495658467885098743394422125448770664780915884607499887124007652170575179788341662562494075890697040002812104276217711177780531531714101170466659914669798731761356006708748071013179523689427521948435305678300228785699782977834784587822891109762500302696156170025046433824377648610283831268330372429267526311653392473167111211588186385133162038400522216579128667529465490681131715993432359734949850904 09

47621322298101726107059611645629909816290555208524790352
40602017279974717534277759277862561943208275051312181562
85512224809394712341451702237358057727861600868838295230
45926478780178899219902707769038953219681986151437803149
97411069260886742962267575605231727775203530353613936

Since the relationship between ϕ and $\frac{1}{\phi}$ is so spectacular, we repeat (for emphasis and to highlight our amazement) that the decimal portion of these two values is the same.

$\frac{1}{\phi}$ = .6180339887498948482045868343656381177203091798057628621

35448622705260462818902449707207204189391137484754088075
38689175212663386222353693179318006076672635443338908659
59395829056383226613199282902678806752087668925017116962
07032221043216269548626296313614438149758701220340805887
95445474924618569536486444924104432077134494704956584678
85098743394422125448770664780915884607499887124007652170
57517978834166256249407589069704000281210427621771117778
05315317141011704666599146697987317613560067087480710131
79523689427521948435305678300228785699782977834784587822
89110976250030269615617002504643382437764861028383126833
03724292675263116533924731671112115881863851331620384005
22216579128667529465490681131715993432359734949850904094
76213222981017261070596116456299098162905552085247903524
06020172799747175342777592778625619432082750513121815628
55122248093947123414517022373580577278616008688382952304
59264787801788992199027077690389532196819861514378031499
74110692608867429622675756052317277752035303613936

It is not periodic,[6] and it is also irrational.[7] The two numbers differ by one. Therefore we get $\frac{1}{\phi} = \phi - 1$. To further solidify the

---

6. A periodic decimal, or repeating decimal, is a decimal that is finite or infinite and has a finite block of digits that eventually repeats indefinitely.
7. An irrational number is a real number that cannot be expressed as an integer or a quotient of integers.

connection between ϕ and the Fibonacci numbers, let's take a short journey back to elementary algebra to solve the equation $\frac{1}{\phi} = \phi - 1$.

If we multiply both sides of the equation by ϕ we get:

$$1 = \phi^2 - \phi.$$

Then $\phi^2 - \phi - 1 = 0$.

Applying the quadratic formula, we get: $\phi = \frac{1 \pm \sqrt{5}}{2}$.

The positive value of

$$\phi = \frac{1 + \sqrt{5}}{2} = 1.6180339887498948482045868343656\ldots.$$

Just to check our earlier presumption (above) that the relationship $\frac{1}{\phi} = \phi - 1$ is really true, we can calculate

$$\frac{1}{\phi} = \frac{2}{\sqrt{5}+1} = \frac{\sqrt{5}-1}{2} \approx 0.6180339887498948482045868343 6564,$$

which appears to bear out our conjecture about the reciprocal relationship.

Therefore, not only does $\phi \cdot \frac{1}{\phi} = 1$ (obviously!), but also

$$\phi - \frac{1}{\phi} = 1.$$

Bear in mind that ϕ and $-\frac{1}{\phi}$ are the roots of the equation $x^2 - x - 1 = 0$, a property we shall examine later.

Powers of the Golden Ratio

It is interesting to examine powers of ϕ. They will further link the Fibonacci numbers to ϕ. To do so we first must find the value of ϕ^2 in terms of ϕ.

$$\phi^2 = \left(\frac{\sqrt{5}+1}{2} \right)^2 = \frac{5 + 2\sqrt{5} + 1}{4} = \frac{2\sqrt{5} + 6}{4} = \frac{\sqrt{5}+3}{2} = \frac{\sqrt{5}+1}{2} + 1 = \phi + 1$$

We now use this relationship ($\phi^2 = \phi + 1$) to inspect the successive powers of ϕ by breaking them down to their component parts in detail. It may at first appear more complicated than it really is. You should try to follow each step (it's really not difficult—and yet very rewarding!) and then extend it to further powers of ϕ.

$$\phi^3 = \phi \cdot \phi^2 = \phi(\phi + 1) = \phi^2 + \phi = (\phi + 1) + \phi = 2\phi + 1$$

$$\phi^4 = \phi^2 \cdot \phi^2 = (\phi + 1)(\phi + 1) = \phi^2 + 2\phi + 1 = (\phi + 1) + 2\phi + 1 = 3\phi + 2$$

$$\phi^5 = \phi^3 \cdot \phi^2 = (2\phi + 1)(\phi + 1) = 2\phi^2 + 3\phi + 1 = 2(\phi + 1) + 3\phi + 1 = 5\phi + 3$$

$$\phi^6 = \phi^3 \cdot \phi^3 = (2\phi + 1)(2\phi + 1) = 4\phi^2 + 4\phi + 1 = 4(\phi + 1) + 4\phi + 1 = 8\phi + 5$$

$$\phi^7 = \phi^4 \cdot \phi^3 = (3\phi + 2)(2\phi + 1) = 6\phi^2 + 7\phi + 2 = 6(\phi + 1) + 7\phi + 2 = 13\phi + 8$$

and so on.

By this point you should be able to see a pattern emerging. As we take further powers of ϕ, the end result of each power of ϕ is actually equal to a multiple of ϕ plus a constant. Further inspection shows that the coefficients of powers of ϕ and the constants are all Fibonacci numbers. Not only that, but they are also in their Fibonacci sequence order. So you ought to be able to extend the list as we have begun in figure 4-2 to get the powers of ϕ:

$$\phi = 1\phi + 0 \qquad\qquad \phi^6 = 8\phi + 5$$
$$\phi^2 = 1\phi + 1 \qquad\qquad \phi^7 = 13\phi + 8$$
$$\phi^3 = 2\phi + 1 \qquad\qquad \phi^8 = 21\phi + 13$$
$$\phi^4 = 3\phi + 2 \qquad\qquad \phi^9 = 34\phi + 21$$
$$\phi^5 = 5\phi + 3 \qquad\qquad \phi^{10} = 55\phi + 34$$
$$\vdots$$

Figure 4-2

Voilà! Once again the Fibonacci number sequence appeared where you may have least expected it. The Fibonacci numbers appear both as the coefficients of ϕ, and as the constants as we take powers of ϕ. Furthermore, we can write all powers of ϕ in a linear form: $\phi^n = a\phi + b$, where a and b are special integers—the Fibonacci numbers.

The Golden Rectangle

For centuries, artists and architects have identified what they believed to be the most perfectly shaped rectangle. This ideal rectangle, often referred to as the "golden rectangle," has also proved to be the most pleasing to the eye. The golden rectangle is one that has the following ratio of its length and width: $\frac{w}{l} = \frac{l}{w+l}$.

The desirability of this rectangle has been borne out by numerous psychological experiments. For example, Gustav Fechner (1801–1887), a German experimental psychologist, inspired by Adolf Zeising's book *Der goldene Schnitt*[8] began a serious inquiry to see if the golden rectangle had a special psychological aesthetic appeal. His findings were published in 1876.[9] Fechner made thousands of measurements of commonly seen rectangles, such as playing cards, writing pads, books, windows, etc. He found that most had a ratio of length to width that was close to ϕ. He also tested people's preferences and found most people preferred the golden rectangle.

What Gustav Fechner actually did was to ask 228 men and 119 women which of the following rectangles is aesthetically the most pleasing. Take a look at the rectangles in figure 4-3. Which rectangle would you choose as the most pleasing to look at? Rectangle 1:1 is too much like a square—considered by the general public as not representative of a "rectangle." It is, after all, a square! On the other hand, rectangle 2:5 (the other extreme) is uncomfortable to look at since it requires the eye to scan it horizontally. Finally, consider the rectangle 21:34, which can be appreciated at a single glance and is therefore more aesthetically pleasing. Fechner's findings seem to bear this out.

8. Adolf Zeising (1810–1876), a German philosopher, *Neue Lehre von den Proportionen des menschlichen Körpers* (New theories about the proportions of the human body) (Leipzig, Germany: R. Weigel, 1854.) The book *Der goldene Schnitt* (The Golden Section) was published postumously (by the Leopoldinisch-Carolinische Akademie: Halle, Germany, 1884).
9. Gustav Theodor Fechner, *Zur experimentalen Ästhetik* (On Experimental Aesthetics) (Leipzig, Germany: Breitkopf & Härtl, 1876).

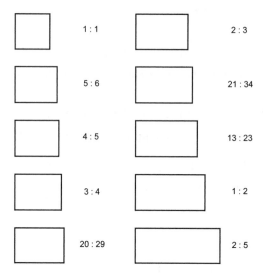

Figure 4-3
Fechner's rectangles.

Here are the results that Fechner reported:

Ratio of sides of rectangle	Percent response for best rectangle	Percent response for worst rectangle
1:1 = 1.00000	3.0	27.8
5:6 = .83333	.2	19.7
4:5 = .80000	2.0	9.4
3:4 = .75000	2.5	2.5
20:29 = .68966	7.7	1.2
2:3 = .66667	20.6	0.4
21:34 = .61765	**35.0**	**0.0**
13:23 = .56522	20.0	0.8
1:2 = .50000	7.5	2.5
2:5 = .40000	1.5	35.7
	100.00	100.00

Figure 4-4

Fechner's experiment has been repeated with variations in methodology many times and his results have been further supported. For example, in 1917 Edward Lee Thorndike (1874–1949), an American psychologist and educator, carried out similar experiments, with analogous results.

In general, the rectangle with the ratio of 21:34 was most preferred. Do those numbers look familiar? Yes, once again the Fibonacci numbers. The ratio $\frac{21}{34} = 0.61764705882352941$ approaches the value of $\frac{1}{\phi}$, and gives us the so-called golden rectangle.

Consider a rectangle (figure 4-5) where the length, l, and the width, w, are in the following proportion: $\frac{w}{l} = \frac{l}{w+l}$.

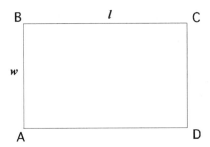

Figure 4-5

By cross multiplying in this proportion we get $w(w + l) = l^2$, or $w^2 + wl = l^2$, or $w^2 + wl - l^2 = 0$.

If we let $l = 1$, then $w^2 + w - 1 = 0$.

Using the quadratic formula,[10] we get $w = \frac{-1 \pm \sqrt{5}}{2}$. Because we are dealing with lengths, the negative value is of no interest here.

Therefore, $w = \frac{-1 + \sqrt{5}}{2} = \frac{\sqrt{5} - 1}{2} = \frac{1}{\phi}$, and the golden ratio again emerges.

---

10. The quadratic formula presented in the high school algebra course, $x = \frac{-b \pm \sqrt{b^2 - 4ac}}{2a}$, is for the general quadratic equation $ax^2 + bx + c = 0$.

This is the same equation that gave us $\frac{1}{\phi}$ above. So we now know that the ratio of the rectangle's dimensions is

$$\frac{w}{l} = \frac{l}{w+l} = \frac{1}{\phi} = \frac{\sqrt{5}-1}{2} \text{, or } \frac{l}{w} = \frac{w+l}{l} = \phi = \frac{\sqrt{5}+1}{2}.$$

giving us a golden rectangle.

Let's see how this rectangle may be constructed using the traditional Euclidean tools: an unmarked straightedge and a pair of compasses. (Another way would be to use a computer geometric-construction program such as Geometer's Sketchpad.) With a width of 1 unit, our objective is to get the length to be $\frac{\sqrt{5}+1}{2}$, so that the ratio of the length to the width will be ϕ, which equals $\frac{\sqrt{5}+1}{2}$.

Perhaps one of the simpler ways to construct this golden rectangle is to begin with a square $ABEF$ (see figure 4-6) with M, the midpoint of $\overline{AF}$. Then with radius $\overline{ME}$ and center M, draw a circle to intersect $\overrightarrow{AF}$ at D. The perpendicular at D intersects $\overrightarrow{BE}$ at C. We now have $ABCD$ (which turns out to be a golden rectangle).

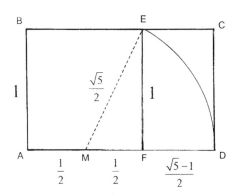

Figure 4-6

Let us verify that figure 4-6 is, in fact, a golden rectangle. Without loss of generality, we let $ABEF$ be a unit square. Therefore $EF = AF = 1$ and $MF = \frac{1}{2}$.

By applying the Pythagorean theorem to ΔMFE,[11] we get $ME = \frac{\sqrt{5}}{2}$. Therefore, $AD = \frac{\sqrt{5}+1}{2}$.

To verify that $ABCD$ is a golden rectangle, we would want to show that the ratio of the lengths and the widths follows the relationship mentioned earlier—namely, that $\frac{CD}{AD} = \frac{AD}{CD+AD}$. We can do this by substituting the above lengths into this proportion to get

$$\frac{1}{\frac{\sqrt{5}+1}{2}} = \frac{\frac{\sqrt{5}+1}{2}}{1+\frac{\sqrt{5}+1}{2}}$$

which is a true equality!

Johannes Kepler (1571–1630), the revered astronomer and mathematician, said that "geometry holds two great treasures: one is the Pythagorean theorem and the other is the golden section. The first we can compare to a bushel of gold and the second we can call a priceless gem."

As we like to relate mathematical phenomena to each other, we shall mention that Underwood Dudley[12] drew a "cute" relationship between the golden ratio and π. He showed that the following is just a good approximation, but nothing more:

$$3.1415926535897932384\ldots = \pi \approx \frac{6}{5}\phi^2 = 3.1416407864998738178\ldots$$

---

11. Applying the Pythagorean theorem gives us:

$$(ME)^2 = (MF)^2 + (EF)^2$$

$$(ME)^2 = \left(\frac{1}{2}\right)^2 + 1^2$$

$$(ME)^2 = \frac{5}{4}$$

$$(ME) = \sqrt{\frac{5}{4}} = \frac{\sqrt{5}}{2}$$

12. *Mathematical Cranks* (Washington, DC: Mathematical Association of America, 1992).

Furthermore, as we continue to connect mathematical values, consider one of the most famous relationships in mathematics. It is attributed to Leonhard Euler (1707–1783), one of the most prolific mathematicians in history. The beauty of this relationship is that the simple equation contains each of the most significant values in mathematics.

It is: $e^{\pi i} + 1 = 0$, where e is the base of natural logarithms (Euler's number), π is the ratio of the circumference of a circle to its diameter (Ludolph's number), i is the imaginary unit of the complex numbers (square root of -1; $\sqrt{-1}$), 1 is the unit of the natural numbers (all numbers > 0 are produced by 1 and adding), and 0 is the neutral element of the addition. Notice that ϕ is missing. Well, we can fix that!

We know that $\phi = \dfrac{\sqrt{5}+1}{2}$ and $1 = -e^{\pi i}$ (from Euler's equation), so we can incorporate ϕ into this equation by replacing the 1 in the first equation with $(-e^{\pi i})$ to get: $\phi = \dfrac{\sqrt{5}-e^{\pi i}}{2}$. So now we have an exact way to relate ϕ to π, e, and i.

Constructing the Golden Section

There are many ways to construct the golden ratio with Euclidean tools (i.e., an unmarked straightedge and a pair of compasses) besides the method shown in figure 4-6. With each method we have an opportunity to see how many lovely geometric relationships are brought into play in our quest for the golden ratio. We shall show a few of them here. The first, shown in figure 4-7, is attributed to Heron of Alexandria (10–70 CE).

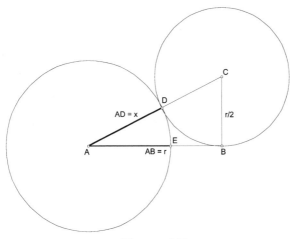

Figure 4-7

You may be able to see from figure 4-7 how the construction is to be done to get $\overline{AB}$ divided into the golden ratio. You begin with $\overline{AB}$ and construct $\overline{AB} \perp \overline{BC}$ ($\perp$ is the symbol used to denote perpendicularity), where $BC = \frac{1}{2} AB$. Then draw the circle with center at C and radius length CB to intersect the hypotenuse of the right triangle ABC at D. You complete the construction by drawing the circle with center A and radius AD to intersect the other leg at E, which is the point on $\overline{AB}$ that divides the segment into the golden section. By using the Pythagorean theorem on triangle ABC, we find that the hypotenuse, $\overline{AC} = \frac{r\sqrt{5}}{2}$. The result is that $\frac{AE}{BE} = \frac{x}{r-x} = \frac{\sqrt{5}+1}{2} = \phi \approx 1.618033988$. This is a start for justifying our conclusion. (Further details and justification can be found in appendix B.)

Another Construction of the Golden Section

Another construction of the golden section along $\overline{AB}$, using Euclidean tools, can be seen in figure 4-8. Here the two circles, with the line segment lengths indicated, are shown to be tangent at D and $\overline{AB} \perp \overline{AC}$.

To do the construction, begin with a right triangle ABC where $AB = a$ and $AC = a/2$. Then draw a circle with center C and radius length CB. Extend $\overrightarrow{CA}$ to meet the circle at D. Next draw the circle with center A and radius length AD. We then can conclude that point E divides the line segment $\overline{AB}$ into the golden ratio. That is, $\frac{AE}{BE} = \frac{x}{a-x} = \frac{\sqrt{5}+1}{2} = \phi \approx 1.618033988$. (The justification for this construction can be found in appendix B.)

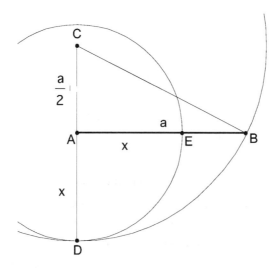

Figure 4-8

Yet Another Construction of the Golden Section

The construction shown in figure 4-9 will have two segments that reflect the golden ratio:

$$\phi = \frac{\sqrt{5}+1}{2} \text{ and } \frac{1}{\phi} = \frac{\sqrt{5}-1}{2}$$

They are $\overline{BQ}$ and $\overline{BP}$.

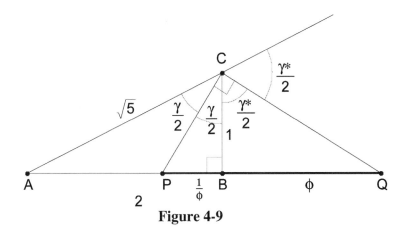

Figure 4-9

The construction is rather simple. You start off with a right triangle ABC, where $AB = 2 \cdot BC$, thus making the hypotenuse $AC = \sqrt{5}$. Then all that remains is to construct the angle bisectors of $\angle ACB$ and the exterior angle at C. (See appendix B.)

A Surprising Construction of the Golden Section

We can construct the golden ratio along a line and then, of course, use the segments to construct the golden rectangle. The construction is simple, as is the justification. Begin with an equilateral triangle ABC inscribed in a circle. Through the midpoint of two of its sides, construct a line that will intersect the circle. (See figure 4-10.) For convenience, we will let the sides of the equilateral triangle be of length 2. It is easy to establish[13] that the line joining the midpoints of the two sides of the triangle is half the third side and, therefore, is of length 1.

---

13. To do this, draw a line segment through E parallel to $\overline{AB}$ and intersecting $\overline{BC}$ at F. The figure formed, $BDEF$ is a rhombus with each side of length 1.

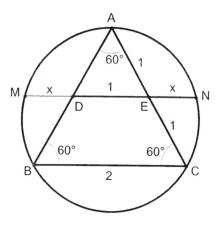

Figure 4-10

Using the relationship that the products of the segments of two intersecting chords of a circle are equal,[14] we get:

$$AE \cdot EC = ME \cdot EN$$
$$1 \cdot 1 = (x + 1)\, x$$
$$x^2 + x - 1 = 0$$
$$x = \frac{\sqrt{5} - 1}{2}, \text{ which is } \frac{1}{\phi}.$$

So, from the previous constructions of the golden section, we can say that the Fibonacci numbers appear even when there is no golden rectangle.

Golden or Fibonacci Spirals

Let us continue our discussion with golden rectangle *ABCD*, but now in a very curious way. We established in figure 4-6 that when a square is constructed internally (as shown in figure 4-11), if $AF = 1$ and $AD = \phi$, then $FD = \phi - 1 = \frac{1}{\phi}$. We now can establish

---

14. When two chords of a circle intersect, the product of the segments of one chord equals the product of the segments of the other chord.

that rectangle $CDFE$ has dimensions $FD = \frac{1}{\phi}$ and $CD = 1$. If we inspect the ratio of length to width of rectangle $CDFE$, we get $\frac{EF}{FD} = \frac{1}{\frac{1}{\phi}} = \phi$, and it is, therefore, also a golden rectangle.

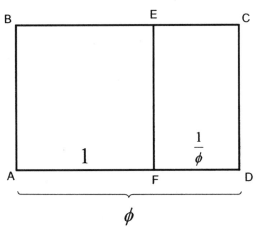

Figure 4-11

Let's continue this process of constructing an internal square in the newly formed golden rectangle. In golden rectangle $CDFE$, square $DFGH$ is constructed (figure 4-12). We find that $CH = 1 - \frac{1}{\phi} = \frac{1}{\phi^2}$, so the ratio of the length to width of rectangle $CHGE$ is

$$\frac{\frac{1}{\phi}}{\frac{1}{\phi^2}} = \phi$$

(having multiplied both numerator and denominator by ϕ^2). This thereby establishes rectangle $CHGE$ also as a golden rectangle.

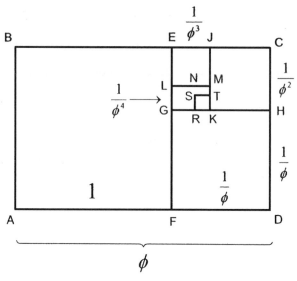

Figure 4-12

Continuing this scheme, we construct square *CHKJ* in golden rectangle *CHGE*. We find[15]

$$EJ = \frac{1}{\phi} - \frac{1}{\phi^2} = \frac{\phi-1}{\phi^2} = \frac{\frac{1}{\phi}}{\phi^2} = \frac{1}{\phi^3}$$

We now inspect the ratio of the dimensions of rectangle *EJKG*. This time the length to width ratio is

$$\frac{\frac{1}{\phi^2}}{\frac{1}{\phi^3}} = \phi$$

Once again, we have a new golden rectangle; this time rectangle *EJKG*.

By continuing this process, we get golden rectangle *GKML*, golden rectangle *NMKR*, golden rectangle *MNST*, and so on.

---

15. We showed earlier that $\phi - \frac{1}{\phi} = 1$, therefore, $\phi - 1 = \frac{1}{\phi}$.

center E, radius EB

center G, radius GF

center K, radius KH

center M, radius MJ

center N, radius NL

center S, radius SR

The result is an approximation of a logarithmic spiral (figure 4-13).

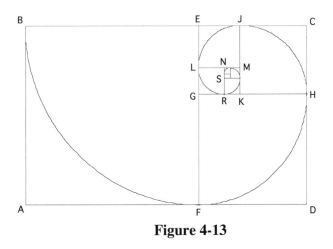

Figure 4-13

The symmetric parts of this complex-looking figure are the squares. Suppose we locate the center of each of these squares. We can draw arcs through each of these points and then see that the centers of these squares lie in another approximation of a logarithmic spiral (figure 4-14).

The spiral in figure 4-13 seems to converge (i.e., end) at a point in rectangle $ABCD$. This point is at the intersection P, of $\overline{AC}$ and $\overline{ED}$ (figure 4-15).

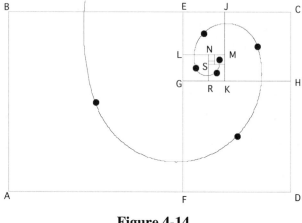

Figure 4-14

Consider once again golden rectangle *ABCD* (figure 4-15). Earlier we established that square *ABEF* determined another golden rectangle *CEFD*.

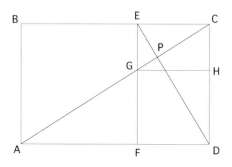

Figure 4-15

Since all golden rectangles have the same shape, rectangle *ABCD* is similar to rectangle *CEFD*. This implies that $\triangle ECD$ is similar to $\triangle CDA$. Therefore $\angle CED$ is congruent to $\angle DCA$. And $\angle DCA$ is complementary to $\angle ECA$. Therefore $\angle CED$ is complementary to $\angle ECA$. Thus $\angle EPC$ must be a right angle, or $\overline{AC} \perp \overline{ED}$.

If the width of one rectangle is the length of the other and the rectangles are similar, then the rectangles are said to be *reciprocal rectangles*. In this case, the ratio of similitude[16] is ϕ.

In figure 4-15 we see that rectangle *ABCD* and rectangle *CEFD* are reciprocal rectangles. Furthermore, we see that reciprocal rectangles have corresponding diagonals that are perpendicular.

In the same way as before, we can prove that rectangles *CEFD* and *CEGH* are reciprocal rectangles. Their diagonals $\overline{ED}$ and $\overline{CG}$ are perpendicular at *P*. This may be extended to each pair of consecutive golden rectangles shown in figure 4-16. Clearly *P* ought to be the limiting point of the spiral.

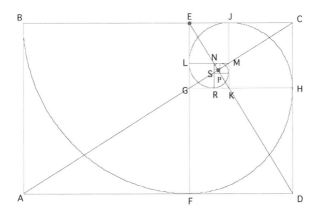

Figure 4-16

We can use this relationship of the diagonals to construct consecutive golden rectangles. We could simply begin with golden rectangle *ABCD* and construct a perpendicular from *D* to $\overline{AC}$ and from its intersection *E* with $\overline{BC}$ construct a perpendicular to $\overline{AD}$ to complete the second golden rectangle. This process can be repeated indefinitely.

Let's take another look at the spiral we generated by drawing quarter circles—one that approximates the golden spiral (see figure 4-17). We have a golden rectangle *ABCD*, with sides of length *a* and *b*

---

16. The ratio of similitude is the ratio of the corresponding sides of the two similar figures, in this case, rectangles.

($a > b$), where it follows that $a = \phi \cdot b$ and $b = \phi^{-1} \cdot a$. We shall then construct the spiral with the quarter circles as before. This will allow us to get the length of the spiral that approximates the golden spiral.

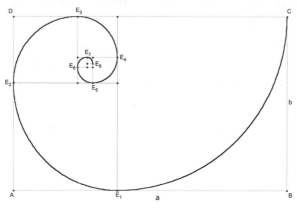

Figure 4-17

The actual golden spiral does not evolve from these quarter circles. This diagram merely gives us a good and easily understood approximation. The table in figure 4-18 provides a progressive calculation of the length of the spiral.

Length	Width (quarter circle radius)			
$a = a_0$	$b = b_0$	$= \phi^{-1} \cdot a$	$= \dfrac{\sqrt{5}-1}{2} \cdot a$	$= \dfrac{1\sqrt{5}-1}{2} \cdot a$
$a_1 = b_0$	$b_1 = a_0 - b_0$	$= \phi^{-1} \cdot a_1$	$= \dfrac{3-\sqrt{5}}{2} \cdot a$	$= -\dfrac{1\sqrt{5}-3}{2} \cdot a$
$a_2 = b_1$	$b_2 = a_1 - b_1$	$= \phi^{-1} \cdot a_2$	$= \dfrac{2\sqrt{5}-4}{2} \cdot a$	$= \dfrac{2\sqrt{5}-4}{2} \cdot a$
$a_3 = b_2$	$b_3 = a_2 - b_2$	$= \phi^{-1} \cdot a_3$	$= \dfrac{7-3\sqrt{5}}{2} \cdot a$	$= -\dfrac{3\sqrt{5}-7}{2} \cdot a$
$a_4 = b_3$	$b_4 = a_3 - b_3$	$= \phi^{-1} \cdot a_4$	$= \dfrac{5\sqrt{5}-11}{2} \cdot a$	$= \dfrac{5\sqrt{5}-11}{2} \cdot a$
$a_5 = b_4$	$b_5 = a_4 - b_4$	$= \phi^{-1} \cdot a_5$	$= \dfrac{18-8\sqrt{5}}{2} \cdot a$	$= -\dfrac{8\sqrt{5}-18}{2} \cdot a$
$a_6 = b_5$	$b_6 = a_5 - b_5$	$= \phi^{-1} \cdot a_6$	$= \dfrac{13\sqrt{5}-29}{2} \cdot a$	$= \dfrac{13\sqrt{5}-29}{2} \cdot a$
$a_7 = b_6$	$b_7 = a_6 - b_6$	$= \phi^{-1} \cdot a_7$	$= \dfrac{47-21\sqrt{5}}{2} \cdot a$	$= -\dfrac{21\sqrt{5}-47}{2} \cdot a$
...				

Figure 4-18

Amazingly, as you inspect the results, you will see the Fibonacci numbers, F_n (1, 1, 2, 3, 5, 8, 13, 21, . . .), as the product with $\sqrt{5}$, as well as the constants highlighting the Lucas numbers, L_n (1, 3, 4, 7, 11, 18, 29, 47, . . .). This could justify calling this spiral the *Fibonacci-Lucas spiral*.

The real golden spiral, also called a logarithmic spiral, looks something like that in figure 4-19.

The real golden (logarithmic) spiral cuts the sides of the squares at very small angles. The apparent spiral (which consists of quarter circles) touches it. So the sides of the golden rectangles aren't tangents to this golden spiral (as in the case of the approximation); they are each cut twice.

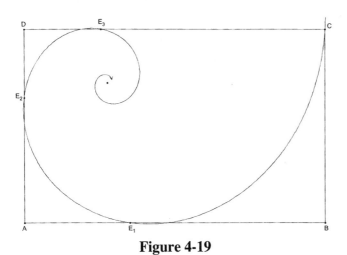

Figure 4-19

It is sometimes called an "equiangular spiral," since the radius keeps a constant angle with the curve. It was named such by René Descartes (1596–1650), the French mathematician, who is responsible for the field of analytic geometry that is done on a "Cartesian plane"—named for its founder. Descartes mentioned this spiral in 1638 correspondences with another French mathematician Marin Mersenne (1588–1648), who is famous for his work with prime numbers. The Swiss mathematician Jacob Bernoulli (1655–1705) referred to this spiral as the logarithmic spiral. He was so enchanted with it and its properties that he requested it as his epitaph with the words: "Eadem mutata resurgo" (It changed me, and yet remain unchanged).[17]

In nature the nautilus shell exhibits such a golden spiral (figure 4-20).

Figure 4-20

Figure 4-21 shows a curve that crosses the x-axis at the Fibonacci numbers.

The spiral part crosses at 1, 2, 5, 13, and so on, on the positive axis, and 0, 1, 3, 8, and so on, on the negative axis.

The oscillatory part crosses at 0, 1, 1, 2, 3, 5, 8, 13, etc. on the positive axis.

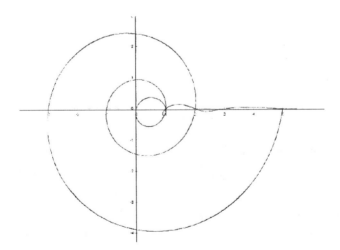

Figure 4-21

---

17. The sculptor who then created the epitaph chiseled not a logarithmic spiral, but rather an Archimedean one!

The curve is strangely reminiscent of the shells of nautilus and snails (figure 4-22). This is not surprising, since the curve tends to a logarithmic spiral as it expands.

Figure 4-22

A Surprising Sighting of the Fibonacci Numbers

We typically call the area between two concentric circles a ring. In figure 4-23, you will notice that as the area of the ring gets smaller, the area of the ellipse that is tangent to each of the circles gets larger. At some point the ellipse will be equal in area to the ring. Strangely enough, the ellipse will be equal in area to the ring[18] when the ratio of the radii of the two circles is 0.618 . . . or $\frac{1}{\phi}$.

---

18. The justification for this claim is as follows: The area of a circle of radius r is πr^2. The area of an ellipse with "radii" a and b (as shown above) is πab. (Note how when $a = b$ in the ellipse, it becomes a circle and the two formulas are the same.) So the outer circle has radius b, the inner circle radius a, and the area of the ring between them is therefore: $\pi (b^2 - a^2)$. This is equal to the area of the ellipse when $\pi (b^2 - a^2) = \pi ab$, and $b^2 - a^2 - ab = 0$.

If we let the *ratio of the two circles' radii* = b/a, be R, say, then dividing the equation by a^2, we have $R^2 - R - 1 = 0$, which means R is ϕ . The equation of an ellipse is $(x/b)^2 + (y/a)^2 = 1$. When $a = b$, we have the equation of a circle of radius $a(=b)$: $(x/a)^2 + (y/a)^2 = 1$.

Again, when you least expect it, ϕ or $\frac{1}{\phi}$, or for that matter, the Fibonacci numbers, come up to greet you.

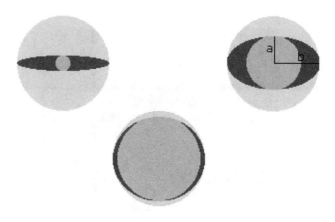

Figure 4-23

Let's look at the right triangle and consider when the tangent of an angle would be equal to the cosine of the same angle—that is, when is $\tan \angle A = \cos \angle A$?

Consider the right triangle ABC (figure 4-24) with $AC = 1$ and $BC = a$. Then by the Pythagorean theorem, we get:
$AB = \sqrt{a^2 + 1}$.

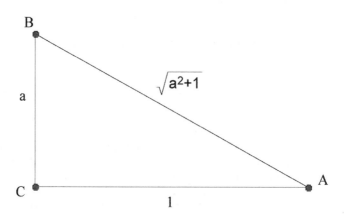

Figure 4-24

$$\tan \angle A = \frac{a}{1}$$

$$\cos \angle A = \frac{1}{\sqrt{a^2+1}}$$

We want to have $\tan \angle A = \cos \angle A$:

$$\frac{a}{1} = \frac{1}{\sqrt{a^2+1}}$$

Solving this equation for a:

$$a\sqrt{a^2+1} = 1$$

$$a^2\left(a^2+1\right) = 1$$

$$a^4 + a^2 - 1 = 0$$

We shall let $p = a^2$, so that we then get the equation $p^2 + p - 1 = 0$, at which point you should recognize this equation.

Then $p = \dfrac{\sqrt{5}-1}{2} = \dfrac{1}{\phi}$, or $a^2 = \dfrac{1}{\phi}$.

Let's use these values for the sides of right triangle ABC (see figure 4-25).

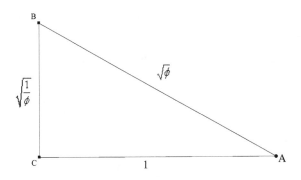

Figure 4-25

So you can see that when a triangle has sides of lengths: $1, \sqrt{a^2+1} = \sqrt{\frac{1}{\phi}+1} = \sqrt{\phi}$, and $a = \sqrt{\frac{1}{\phi}}$, the tangent of an acute angle of the right triangle will equal the cosine. Again, the golden ratio—or, if you wish, the Fibonacci numbers—appears when you least expect it.

Another Emergence of the Fibonacci Numbers in Geometry

An engaging problem that was posed by J. A. H. Hunter[19] asks us to determine the points on any rectangle $ABCD$ that will leave three triangles of equal areas when the center triangle (triangle 4 in figure 4-26) is removed from it.

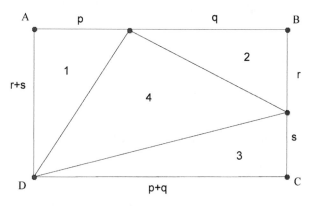

Figure 4-26

You can guess by now that this will in some way result in the Fibonacci numbers—another sighting in geometry of this wonderful sequence. We begin (using the markings in figure 4-26) by setting the areas of the three triangles equal to one another.

$$Area\Delta 1 = \frac{p(r+s)}{2}, \quad Area\Delta 2 = \frac{qr}{2}, \quad Area\Delta 3 = \frac{s(p+q)}{2}$$

Since $Area\Delta 1 = Area\Delta 2 = Area\Delta 3,$

$$\frac{p(r+s)}{2} = \frac{qr}{2}, \quad \text{and} \quad \frac{p(r+s)}{2} = \frac{s(p+q)}{2}, \text{ or}$$

$$p(r+s) = qr, \quad \text{and} \quad p(r+s) = s(p+q)$$

$$p = \frac{qr}{r+s} \quad \text{and} \quad pr = sq$$

---

19. "Triangle Inscribed in a Rectangle," *Fibonacci Quarterly* 1 (1963): 66.

From the second equation we can set up the following propor-
tion: $\frac{p}{s} = \frac{q}{r}$, which tells us that the two sides of the rectangle were
divided proportionally. But what is that proportion? (Can you
guess by now?)

Let us replace p from the second equation, above, with the
value of p shown in the first equation:

$$\left(\frac{qr}{r+s}\right)r = sq$$

$$r^2 = s\left(r+s\right)$$

If we divide both sides of the equation by s^2, we will get:

$$\frac{r^2}{s^2} = \frac{s\left(r+s\right)}{s^2}$$

$$\frac{r^2}{s^2} = \frac{sr}{s^2} + \frac{s^2}{s^2}$$

$$\frac{r^2}{s^2} = \frac{r}{s} + 1$$

Putting this in a more recognizable form (remember, ϕ and $\frac{1}{\phi}$
are the roots of the equation $x^2 - x - 1 = 0$):

$$\left(\frac{r}{s}\right)^2 - \frac{r}{s} - 1 = 0$$

We can see (because $r > s$, $q > p$) that the solution of this
equation is $\frac{r}{s} = \frac{1+\sqrt{5}}{2} = \phi$ ($= \frac{q}{p}$).

Therefore, the points on the sides of the rectangle that will determine the triangles that must be removed to leave the three triangle of equal areas must be placed so that they partition the sides in
the golden ratio (ϕ). Again, the Fibonacci numbers are involved!

The Diagonal of the Golden Rectangle

We have done quite a bit with the golden rectangle, yet there never seems to be an end to what you can do. For example, the golden rectangle—the relationship of whose sides are in the golden ratio—allows us a neat way to find the point along the diagonal that cuts it into a ratio related to the golden ratio. It is only because of the unique properties of this special rectangle that we can do this so easily.

Consider the golden rectangle $ABCD$, whose sides $AB = a$ and $BC = b$, so that $\frac{a}{b} = \phi$. As shown in figure 4-27, two semicircles are drawn on the sides $\overline{AB}$ and $\overline{BC}$ to intersect at S. If we now draw segments $\overline{SA}$, $\overline{SB}$, and $\overline{SC}$, we find that $\angle ASB$ and $\angle BSC$ are right angles (since they are each inscribed in a semicircle). Therefore, $\overline{AC}$ is a straight line, namely, the diagonal. We can now show, rather elegantly, that point S divides the diagonal in the square of the golden ratio.

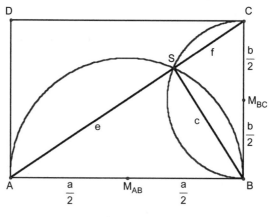

Figure 4-27

From the mean proportional ratios (obtained from similar triangles $\triangle ABC$, $\triangle ASB$, and $\triangle BSC$), we get the following:

We use the segment length markings in figure 4-27, for each of the right triangles:

$\triangle ASB: \quad a^2 = e(e+f)$

$\triangle BSC: \quad b^2 = f(e+f)$

Therefore, $\dfrac{a^2}{b^2} = \dfrac{e}{f}$.

But since $\dfrac{a}{b} = \phi$, then $\dfrac{a^2}{b^2} = \phi^2 = \phi + 1$

[Recall: $\phi = \dfrac{1}{\phi} + 1$ and the powers of ϕ.]

Thus the point S divides the diagonal of the golden rectangle in a ratio involving the golden section $\left(\dfrac{\phi^2}{1} \text{ or } \dfrac{\phi+1}{1}\right)$ and, in turn, the Fibonacci numbers.

Another Curiosity That Will Generate the Golden Ratio

Consider the semicircle in figure 4-28 with the three congruent circles inscribed so that the tangency points are as shown.

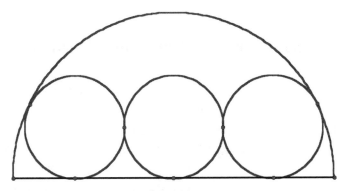

Figure 4-28

We seek to find the ratio of the radii of the large semicircle to one of the smaller circles. In figure 4-29, $AB = 2R$ and $AM = R$. Each of the congruent small circles has a radius r. Consider right triangle *CKM* with legs *r* and 2*r*, where the hypotenuse then has length $r\sqrt{5}$. So we now have $MK = r\sqrt{5}$ and $PK = r$, and therefore, $MP = r(\sqrt{5}+1) = R$.

Put another way, $\dfrac{R}{r} = \sqrt{5}+1 = 2\phi$.

Out of this seemingly unrelated situation with three congruent circles inscribed in a semicircle, we find the ratio of their radii is related to the golden ratio.

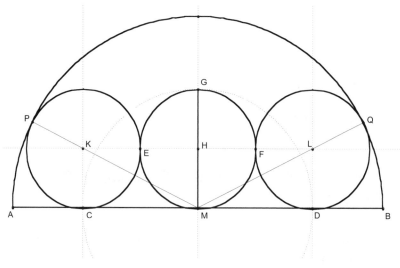

Figure 4-29

Fibonacci Numbers and a Curious Dilemma

This discussion of the role of the Fibonacci numbers in geometry can also take on other entertaining aspects. Toward this end, we will examine a rather curious problem. It was made popular by the English mathematician Charles Lutwidge Dodgson (1832–1898), who, under the pen name of Lewis Carroll, wrote *The Adventures of Alice in Wonderland*.[20] He posed the following problem: the square on the left side of figure 4-30 has an area of 64 square units and is partitioned as shown.

---

20. See Stuart Dodgson Collingwood, ed., *Diversions and Digressions of Lewis Carrol* (New York: Dover, 1961), pp. 316–17.

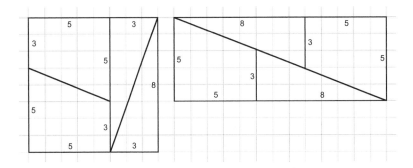

Figure 4-30

These four parts are then reassembled to form the rectangle at the right of figure 4-30. This rectangle has an area of $13 \cdot 5 = 65$ square units. Where did this additional square unit come from? Think about it before reading further.

All right, we'll relieve you of the suspense. The "error" lies in the assumption that the figures will all line up along the drawn diagonal. This, it turns out, is not so. In fact, a "narrow" parallelogram is embedded here, and it has an area of one square unit (see figure 4-31).

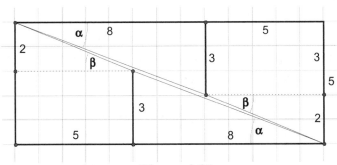

Figure 4-31

We can discover where the error lies by taking the tangent function of α and β so that we can discover the measure of these angles. Remember, they ought to be equal if they lie on the diagonal.[21]

---

21. Alternate-interior angles of parallel lines are congruent.

Since $\tan \alpha = \dfrac{3}{8}$, then $\alpha \approx 20.6°$.

Since $\tan \beta = \dfrac{2}{5}$, then $\beta \approx 21.8°$.

The difference, $\beta - \alpha$, is merely $1.2°$, yet enough to show that they are not on the diagonal.

You will notice that the segments above were 2, 3, 5, 8, and 13 —all Fibonacci numbers. Moreover, we already discovered[22] that $F_{n-1}F_{n+1} = F_n^2 + (-1)^n$, where $n \geq 1$. The rectangle has dimensions 5 and 13, and the square has a side length 8. These are the fifty, sixth, and senenth Fibonacci numbers: $F_5, F_6,$ and F_7 .

This relationship tells us that

$$F_5 F_7 = F_6^2 + (-1)^6$$

$$5 \cdot 13 = 8^2 + 1$$

$$65 = 64 + 1$$

This puzzle can then be done with any three consecutive Fibonacci numbers as long as the middle number is an even-numbered member of the Fibonacci sequence (i.e., in an even position). If we use larger Fibonacci numbers, the parallelogram will be even less noticeable. But if we use smaller Fibonacci numbers, then our eyes cannot be deceived as in figure 4-32:

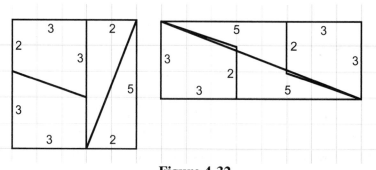

Figure 4-32

22. Chapter 1, page 55, item 11.

Here is the general form of the rectangle (figure 4-33).

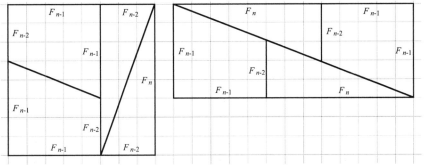

Figure 4-33

To do this properly, without the missing area, the only partitioning—amazingly enough—is with the golden ratio, ϕ as is seen in figure 4-34.

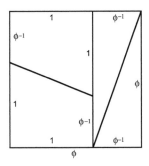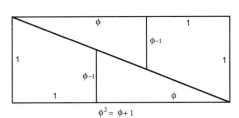

Figure 4-34

The areas of the rectangle and the square are equal here (figure 4-34), as we shall show as follows:

The area of the square

$$= \phi \cdot \phi = \phi^2 = \phi + 1 = \frac{\sqrt{5} + 3}{2} = 2.6180339887 \ldots$$

The area of the rectangle

$$= (\phi + 1) \cdot 1 = \phi + 1 = \frac{\sqrt{5} + 3}{2} = 2.6180339887 \ldots$$

Thus the areas of the square and the rectangle under this partition are equal.

The Golden Triangle

Now that we have thoroughly investigated the famous golden rectangle, we are ready to consider the golden ratio as it pertains to a triangle—the golden triangle. As you would expect, much like the golden rectangle, which has the Fibonacci numbers embedded within it, so, too, does this golden triangle exhibit the Fibonacci numbers, and consequently, the golden ratio. Let's consider a triangle that contains this golden ratio. We will begin with an attempt to place an isosceles triangle into another similar isosceles triangle in a somewhat analogous way to the way we embedded our similar golden rectangles earlier. To do this, we may get a configuration as in figure 4-35. The sum of the measures of the angles of triangle ABC is $\alpha + \alpha + \alpha + 2\alpha = 5\alpha = 180°$, and therefore $\alpha = 36°$.

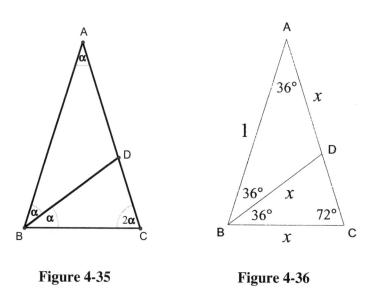

Figure 4-35   Figure 4-36

Aside from placing the two similar triangles the way we did above, we could also have simply begun with an isosceles triangle whose vertex angle measures 36°. Construct the bisector $\overline{BD}$ of $\angle ABC$ (figure 4-36).

Since isosceles triangles with congruent vertex angles are similar, then $\triangle ABC$ is similar to $\triangle BCD$. We shall now let $AD = x$

and $AB = 1$. However since $\triangle ADB$ and $\triangle DBC$ are isosceles, $BC = BD = AD = x$.

From the similarity above, we get a (hopefully by now) familiar equation: $\dfrac{1}{x} = \dfrac{x}{1-x}$.

As before, this gives us: $x^2 + x - 1 = 0$ and $x = \dfrac{\sqrt{5}-1}{2}$. (The negative root cannot be used for the length of $\overline{AD}$.)

Remember that $\dfrac{\sqrt{5}-1}{2} = \dfrac{1}{\phi}$.

In $\triangle ABC$ the ratio of $\dfrac{\text{side}}{\text{base}} = \dfrac{1}{x} = \phi$.

We therefore call this a *golden triangle*. One easy way to construct a golden triangle is to first construct the golden section (done earlier in this chapter; for example, in figure 4-9: $AB = 2$ and $BC = 1$, angle bisectors provide the points P and Q with $BP = \dfrac{1}{\phi}$ and $BQ = \phi$). We draw a circle around O with radius 1. On this circle we choose a point A and draw around this center a second circle with radius $x = \dfrac{1}{\phi}$. The intersection point(s) of the two circles, as shown in figure 4-37, helps to determine a golden triangle (compare also with figure 4-36).

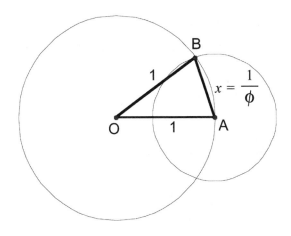

Figure 4-37

By taking consecutive angle bisectors $\overline{BD}$, $\overline{CE}$, $\overline{DF}$, $\overline{EG}$, and $\overline{FH}$ of a base angle of each newly formed 36°, 72°, 72° triangle, we get a series of golden triangles (see figure 4-38). These golden triangles (36°, 72°, 72°) are: $\triangle ABC$, $\triangle BCD$, $\triangle CDE$, $\triangle DEF$, $\triangle EFG$, and $\triangle FGH$. Obviously, had space permitted, we could have continued to draw angle bisectors and thereby generate more golden triangles. Our inspection of the golden triangle will parallel that of the golden rectangle. We proceed in pursuit of the Fibonacci numbers.

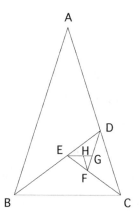

Figure 4-38

Let us begin by having $HG = 1$ (figure 4-38). Since the ratio of $\frac{\text{side}}{\text{base}}$ of a golden triangle is ϕ, we find that for golden $\triangle FGH$:

$$\frac{GF}{HG} = \frac{\phi}{1}, \text{ or } \frac{GF}{1} = \frac{\phi}{1}, \text{ and } GF = \phi.$$

Similarly for golden $\triangle EFG$: $\frac{FE}{GF} = \frac{\phi}{1}$, but $GF = \phi$, so $FE = \phi^2$.

In golden $\triangle DEF$: $\frac{ED}{FE} = \frac{\phi}{1}$, but $FE = \phi^2$, therefore $ED = \phi^3$.

Again, for $\triangle CDE$: $\frac{DC}{ED} = \frac{\phi}{1}$, but $ED = \phi^3$, therefore $DC = \phi^4$.

For ΔBCD: $\dfrac{CB}{DC} = \dfrac{\phi}{1}$, but $DC = \phi^4$, therefore $CB = \phi^5$.

Finally for ΔABC: $\dfrac{BA}{CB} = \dfrac{\phi}{1}$, but $CB = \phi^5$, therefore $BA = \phi^6$.

This can be summarized by using our knowledge of powers of ϕ (developed earlier) as follows (this time we point out the Fibonacci numbers):

$$HG = \phi^0 = 0\phi + 1 = F_0\phi + F_{-1}$$

$$GF = \phi^1 = 1\phi + 0 = F_1\phi + F_0$$

$$FE = \phi^2 = 1\phi + 1 = F_2\phi + F_1$$

$$ED = \phi^3 = 2\phi + 1 = F_3\phi + F_2$$

$$DC = \phi^4 = 3\phi + 2 = F_4\phi + F_3$$

$$CB = \phi^5 = 5\phi + 3 = F_5\phi + F_4$$

$$BA = \phi^6 = 8\phi + 5 = F_6\phi + F_5$$

As we did with the golden rectangle, we can generate an approximation of a logarithmic spiral by drawing arcs to join the vertex angle vertices of consecutive golden triangles (see figure 4-39).

That is, we draw circular arcs as follows:

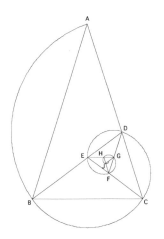

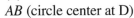

$\overparen{AB}$ (circle center at D)

$\overparen{BC}$ (circle center at E)

$\overparen{CD}$ (circle center at F)

$\overparen{DE}$ (circle center at G)

$\overparen{EF}$ (circle center at H)

$\overparen{FG}$ (circle center at J)

Figure 4-39

There are many other truly fascinating relationships emanating from the golden ratio. After you have been exposed to the golden triangle, the next logical place to turn for more applications is the regular pentagon[23] and the regular pentagram (the five-pointed star), since these are essentially composed of many golden triangles. You will then see that the golden ratio abounds throughout—as do the Fibonacci numbers.

The Golden Angle

Let us first take a side step and look at the golden angle, one that divides a complete circle, $360°$, into the golden ratio. Notice in figure 4-40 the angles ψ and φ are in the ratio that approaches the golden ratio:

$$\psi = 360° - \frac{360°}{\phi} = 137.5077640\ldots° \approx 137.5°.$$

$$\varphi = \frac{360°}{\phi} = 222.4922359\ldots° \approx 222.5°.$$

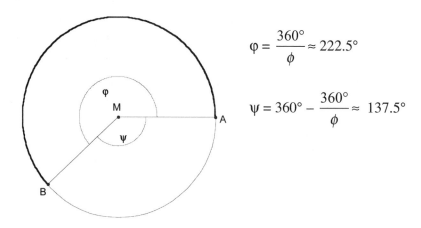

$$\varphi = \frac{360°}{\phi} \approx 222.5°$$

$$\psi = 360° - \frac{360°}{\phi} \approx 137.5°$$

Figure 4-40

---

23. A regular pentagon is one whose sides have the same length and whose angles are all the same measure.

The ratio of $\dfrac{\text{circle}}{\text{larger angle}} = \dfrac{360°}{222.5°} \approx 1.618$, and $\dfrac{\text{larger angle}}{\text{smaller angle}} \approx 1.618$.

In each case, the golden ratio is approximated, as it can be, by the quotient of successive Fibonacci numbers.

The Pentagon and the Pentagram

We now come to the beautiful geometric shape that sums up much of the golden ratio in one configuration.

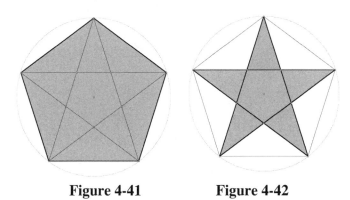

Figure 4-41   **Figure 4-42**

The golden triangle is embedded many times in the regular pentagram, which just happened to be the symbol of the Pythagoreans. According to Pythagoras, all geometric shapes could be described in terms of integers. So it would have come as a great disappointment to him when one of his followers, Hippasus of Metapontum (ca. 450 BCE), showed that the ratio of a diagonal of the regular pentagon (i.e., a side of the regular pentagram) to the side length of the pentagon could not be expressed as a fraction of integers, in other words, this ratio is not rational! This, then, carried over to their symbol: the pentagram. This secret society was a bit troubled by this—which today can be seen as the very beginning of our concept of irrational numbers. That is, these are numbers that cannot be repressed as a ratio of two whole numbers; hence the name *irrational*. In the regular pentagon, the ratio of the side to the diagonal is irrational. But which irrational number did he find? Yes, you guessed it! It was the golden ratio, ϕ.

To show that this length relationship is actually irrational, we use the relationship that in a regular pentagon every diagonal is parallel to the sides it does not intersect. In figure 4-43 the triangles *AED* and *BTC* have parallel sides, so they are similar to each other.

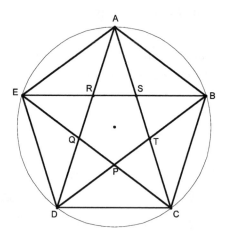

Figure 4-43

Therefore $\frac{AD}{AE} = \frac{BC}{BT}$. But $BT = BD - TD = BD - AE$. In the regular pentagon, therefore, the following ratio holds:

diagonal : side = side : (diagonal – side).

Symbolically, we can write this as

$$\frac{d}{a} = \frac{a}{d-a} \quad \text{or} \quad \frac{d}{a} = \frac{1}{\frac{d}{a}-1}$$

(with *d* as the length of the diagonal and *a* the length of the side).

If we now let $x = \frac{d}{a}$ we get the equation $x = \frac{1}{x-1}$, which then can be converted to the quadratic equation $x^2 - x - 1 = 0$, of which $\frac{d}{a}$ is a positive root and just happens to be the irrational number $\phi = \frac{\sqrt{5}+1}{2}$. (Remember: $\sqrt{5}$ is irrational!)

This is what we claimed at the outset: the ratio of the diagonal to the side of a regular pentagon is irrational. As the irrational number $\pi = 3.1415926535897932384 \ldots$ is connected inseparably with the circle, so, too, the irrational number $\phi = 1.6180339887$ $498948482 \ldots$ is connected inseparably with the regular pentagon!

The regular pentagon $ABCDE$ (figure 4-43) is a fascinating figure with lots of useful properties. Here are some for you to appreciate and perhaps ponder over. You might look for other such properties.

For figure 4-43 the regular pentagon $ABCE$ has the following properties:

(a) The size of every interior angle is $108°$:

$$\angle EAB = \angle ABC = \angle BCD = \angle CDE = \angle DEA = 108°$$

(b) The size of the angles of a golden rectangle are:

$$\angle BEA = \angle CAB = \angle DBC = \angle ECD = \angle ADE = 36°$$
$$\angle PEB = \angle QAC = \angle RBD = \angle SCE = \angle TDA = 36°$$
$$\angle CDA = \angle DEB = \angle EAC = \angle ABD = \angle BCE = 72°$$

(c) The following triangles are all isosceles:

$$\triangle DAC, \triangle EBD, \triangle ACE, \triangle BDA, \text{ and } \triangle CEB$$
$$\triangle BEA, \triangle CAB, \triangle DBC, \triangle ECD, \text{ and } \triangle ADE$$
$$\triangle PEB, \triangle QAC, \triangle RBD, \triangle SCE, \text{ and } \triangle TDA$$

(d) The triangles $\triangle DAC$ and $\triangle QCD$ are similar (as are many others figure 4-43).

(e) All diagonals of the pentagon have the same length.

(f) Every side of the pentagon is parallel to the diagonal "facing" it.

(g) As an example, $AD : DC = CQ : QD$.

(h) The intersection point of two diagonals partitions both diagonals in the golden section.

(i) *PQRST* is a regular pentagon.

Which of the triangles are golden triangles in the regular pentagon?

At this point we can bring in our Fibonacci numbers. We already established that the various line segments of a regular pentagon cannot be measured in integers. We say the segment relationships are incommensurable. Yet we know that the ratio of adjacent Fibonacci numbers approaches the golden ratio. With the obvious loss of accuracy, we can display the regular pentagon and pentagram in terms of the Fibonacci numbers (see figure 4-44).

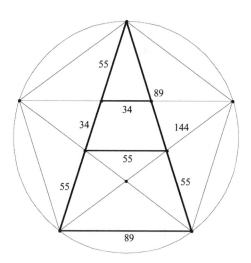

Figure 4-44

These conditions can be approximately interpreted as follows:

If the side and the diagonal of the little pentagon were approximately of lengths 34 and 55 mm, the side length of the larger pentagon (89 mm) might have been obtained from the sum of these values (34 + 55). Similarly, we can generate the other lengths as shown in figure 4-44. There again, we have the Fibonacci numbers!

Remember, the irrationality Hippasus of Metapontum encountered, which made the Pythagoreans a bit disturbed, was that, for example, $\frac{144}{89}$ and $\frac{55}{34}$ have a value of about 1.618. This is only an approximation of the actual irrational value. Had the line segments been exactly 34, 55, 89, and 144 (mm), then all the fame Hippasus of Metapontum garnered would have been ignored. So he remains the "originator" to have stumbled on irrational numbers. Yet all this fame did not come without his attackers. A remark by the Greek philosopher Plato (427–348/347 BCE) shows how terribly the discovery of Hippasus moved the Greek:

> I thought such an ignorance befits not people but rather a herd's pigs, and I am not only ashamed of me but for all Greeks.[24]

The pentagram (five-corner star) inscribed in a regular pentagon abounds with the golden ratio ϕ, since it is composed of lots of golden triangles. In figure 4-45 the regular pentagon has side length 1.

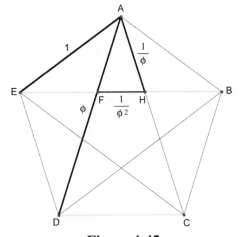

Figure 4-45

---

24. Plato, *Laws for an Ideal State*.

Since the golden triangle is everywhere to be seen in the figure, we have for each golden triangle: $\frac{\text{side}}{\text{base}} = \phi$, therefore $\frac{AD}{DC} = \phi$. With $DC = 1$, we get $AD = \phi$. In the golden $\triangle AEH$ we get

$$\frac{\text{side}}{\text{base}} = \phi = \frac{AE}{AH}.$$

Therefore $AH = \frac{1}{\phi}$.

Since $EH = DC = 1$, then $FH = EH - EF = 1 - \frac{1}{\phi} = \frac{1}{\phi^2}$. [25]

The golden ratio is also involved with various area comparisons in figure 4-45.

The ratio of the area of the larger pentagon $ABCDE$ to the area of the smaller pentagon is $\frac{\phi^4}{1}$.

The ratio of the area of the larger pentagon $ABCDE$ to the area of the pentagram is $\frac{\phi^3}{2}$.

You can see the continued pattern in pentagrams and pentagons (figure 4-46) harboring the golden ratio and, consequently, the Fibonacci numbers.

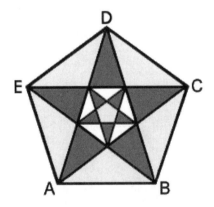

Figure 4-46

---

25. This comes from the now-familiar equation: $\phi^2 - \phi - 1 = 0$, and then we divide both sides by ϕ^2 to get: $1 - \frac{1}{\phi} = \frac{1}{\phi^2}$. See also figure 4-12.

Constructing a Regular Pentagon

The construction of a regular pentagon is more complicated than most other constructible regular polygons. The regular hexagon is easily constructible. One need only draw circles around the circumference of a given circle with the same radius as the given circle. As you can see in figure 4-47, only four such circles are needed. Joining the intersection points gives us a regular hexagon.

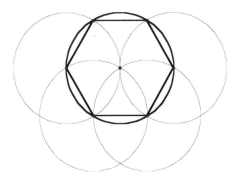

Figure 4-47

Were we to try to construct a regular pentagon in a similar way, we would find ourselves in a dilemma. Perhaps the most important artist Germany has contributed to Western culture is Albrecht Dürer (1471–1528). A much forgotten work by this artist (in 1525) is a geometric construction (using a straightedge and a pair of compasses) of a regular pentagon, which he knew was an approximation of a regular pentagon but extremely close to perfect. It is so nearly perfect that its inaccuracy is not visually detectable. He offered this very easy construction to the mathematical community as an alternative method to get a "regular" pentagon, with the full knowledge that is was off by about half a degree.[26] Its deviation from a perfect regular pentagon is minuscule, but that cannot be

26. C. J. Scriba and P. Schreiber, *5000 Jahre Geometrie. Geschichte, Kulturen, Menschen* (Berlin: Springer, 2000), pp. 259, 289–90.

ignored. Until recently, engineering books still provided it as a method for constructing a regular pentagon. We shall provide an explanation of the method here, despite its being slightly inaccurate, since it is instructive and has been used for many years.

In figure 4-48 we begin with a segment AB. Five circles of radius AB are constructed as follows:

1. Circles with centers at A and B are drawn and intersect at Q and N.
2. Then the circle with center Q is drawn to intersect circles with centers at A and B at points R and S, respectively.
3. $\overline{QN}$ intersects circle Q with center at at P.
4. $\overline{SP}$ and $\overline{RP}$ intersect circles with centers at A and B at points E and C, respectively.
5. Draw the circles with centers at E and C, with radius AB to intersect at D.[27]
6. The polygon $ABCDE$ is (approximately) a regular pentagon.

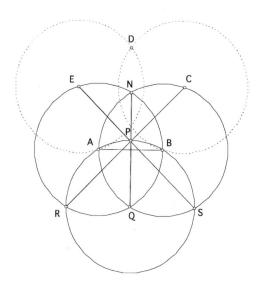

Figure 4-48

<hr>

27. Notice we only need one of the points of intersection (D) of these two circles.

Joining the points in order (figure 4-49), we get the pentagon *ABCDE*.

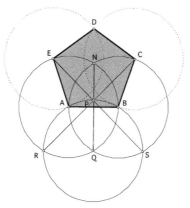

Figure 4-49

Although the pentagon "looks" regular, its $m\angle ABC$ is about $\frac{22}{60}$ of a degree too large. In other words, for *ABCDE* to be a regular pentagon, each angle must be $108°$. Instead we will show that $m\angle ABC \approx 108.3661202°$.

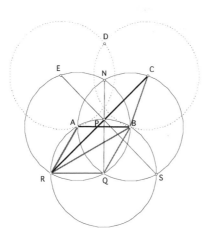

Figure 4-50

The rhombus in figure 4-50, *ABQR*, has $\angle ARQ = 60°$ and $BR = AB\sqrt{3}$ since $\overline{BR}$ is actually twice the length of an altitude of equilateral $\triangle ARQ$. Since $\triangle PRQ$ is an isosceles right triangle, $m\angle PRQ = 45°$, and then $m\angle BRC = 15°$.

We shall apply the law of sines to $\triangle BCR$: $\dfrac{BR}{\sin \angle BCR} = \dfrac{BC}{\sin \angle BRC}$;

that is, $\dfrac{AB\sqrt{3}}{\sin \angle BCR} = \dfrac{AB}{\sin 15°}$, or $\sin \angle BCR = \sqrt{3}\sin 15°$.

Therefore, $m\angle BCR \approx 26.63387984$.

In $\triangle BCR$,

$$m\angle RBC = 180° - m\angle BRC - m\angle BCR$$
$$\approx 180° - 15° - 26.63387984°$$
$$\approx 138.3661202°$$

Thus, since $m\angle ABR = 30°$,

$$m\angle RBC - m\angle ABR \approx 138.661202° - 30° \approx 108.3661202°$$

and *not* $108°$, as it should be in order for it to be a regular pentagon!

The results of Dürer's construction:

$$m\angle ABC = m\angle BAE \approx 108.37°, \ m\angle BCD = m\angle AED$$
$$\approx 107.94°, \text{ and } m\angle EDC \approx 107.38°.$$

To construct a proper regular pentagon, we would first construct a golden triangle and then simply mark its base off along a given circle as shown in figure 4-51.

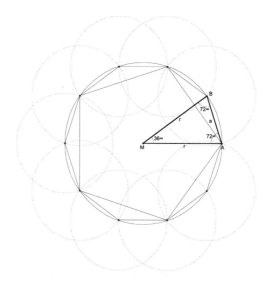

Figure 4-51

We can even create the golden ratio by paper-folding methods. Take a strip of paper, making sure the sides are parallel, and fold it into a regular knot. Then carefully pull it taut so that it looks like picture 3 in figure 4-52. You can tear off the end flops as in picture 4, if you wish. You have now formed a regular pentagon. If you hold this pentagon up to the light and look through the knot, you should see the pentagram inside the pentagon that you formed.

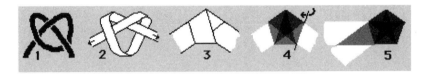

Figure 4-52

We have shown the many geometric sightings where—when you least expect it—the Fibonacci numbers appear and sometimes in the form of the golden ratio. We leave you in this chapter with a series of regular pentagons that are placed so the vertex of each divides the side of another in the golden ratio (see figure 4-53).

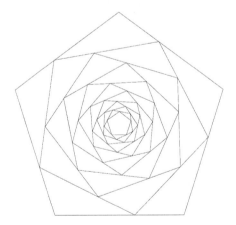

Figure 4-53

If we join specific points (figure 4-54). we can generate a series of spirals reminiscent of those we encountered in chapter 2.

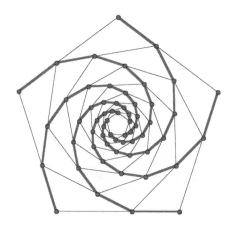

Figure 4-54

In figure 4-55 the sides of neighboring (or adjacent) pentagons are in the ratio equal to $\phi : 1$. Once again, we can use the Fibonacci numbers to render the appropriate approximation.

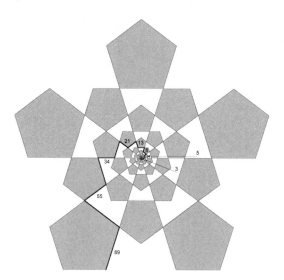

Figure 4-55

There are many more such appearances in geometry that we will leave the reader to discover.

Chapter 5

The Fibonacci Numbers and Continued Fractions

We have seen the Fibonacci numbers as they appear in nature, as they manifest themselves among other numbers and sequences, and as they can be seen in geometry. They also can be seen in the study of continued fractions. A continued fraction is just another form of a fraction that allows us to look at the nature of numbers in a different way. We will introduce you to continued fractions. Once again, after you are comfortable with continued fractions, we will investigate their relationship to the Fibonacci numbers.

Continued Fractions

We will begin with a brief introduction to continued fractions. A continued fraction is a fraction in which the denominator has a mixed number (a whole number and a proper fraction) in it. We can take an improper fraction such as $\frac{13}{7}$ and express it as a mixed number:

$$1\frac{6}{7} = 1 + \frac{6}{7}$$

Without changing the value, we could then write this as

$$1+\frac{6}{7}=1+\cfrac{1}{\cfrac{7}{6}}$$

which in turn could be written (again, without any value change) as:

$$1+\cfrac{1}{1+\cfrac{1}{6}}$$

This is a continued fraction. We could have continued this process, but when we reach a unit fraction (i.e., a fraction in which the numerator is 1 and the denominator is a positive integer; in this case, the unit fraction is $\frac{1}{6}$), we are essentially finished.

So that you can get a better grasp of this technique, we will create another continued fraction. We will convert $\frac{12}{7}$ to continued fraction form. Notice that at each stage, when a proper fraction is reached, we take the reciprocal of the reciprocal (e.g., change $\frac{2}{5}$ to $\cfrac{1}{\cfrac{5}{2}}$, as we will do in the example that follows), which does not change its value:

$$\frac{12}{7}=1+\frac{5}{7}=1+\cfrac{1}{\cfrac{7}{5}}=1+\cfrac{1}{1+\cfrac{2}{5}}=1+\cfrac{1}{1+\cfrac{1}{\cfrac{5}{2}}}=1+\cfrac{1}{1+\cfrac{1}{2+\cfrac{1}{2}}}$$

If we break up a continued fraction into its component parts (called convergents),[1] we get closer and closer to the actual value of the original fraction.

---

1. This is done by considering the value of each portion of the continued fraction up to each plus sign, successively.

First convergent of $\dfrac{12}{7}$ $= 1$

Second convergent of $\dfrac{12}{7}$ $= 1 + \dfrac{1}{1} = 2$

Third convergent of $\dfrac{12}{7}$ $= 1 + \cfrac{1}{1 + \cfrac{1}{2}} = 1 + \dfrac{2}{3} = 1\dfrac{2}{3} = \dfrac{5}{3}$

Fourth convergent of $\dfrac{12}{7}$ $= 1 + \cfrac{1}{1 + \cfrac{1}{2 + \cfrac{1}{2}}} = \dfrac{12}{7}$

The above examples are all *finite* continued fractions, which are equivalent to rational numbers (those that can be expressed as simple fractions). It would then follow that an irrational number would result in an *infinite* continued fraction. That is exactly the case. A simple example of an infinite continued fraction is that of $\sqrt{2}$. Although we show it here, we will actually generate it just a bit further on.

$$\sqrt{2} = 1 + \cfrac{1}{2 + \cfrac{1}{2 + \cfrac{1}{2 + \cfrac{1}{2 + \cfrac{1}{2 + \cfrac{1}{2 + \cdots}}}}}}$$

We have a short way to write a long (in this case infinitely long!) continued fraction: $[1;2,2,2,2,2,2,2, \ldots]$, or when there are these endless repetitions, we can even write it in a shorter form as

$\left[1;\overline{2}\right]$, where the bar over the 2 indicates that the 2 repeats end-lessly.

In general, we can represent a continued fraction as:

$$a_0 + \cfrac{1}{a_1 + \cfrac{1}{a_2 + \cfrac{1}{a_3 + \dots \cfrac{}{} \cfrac{1}{a_{n-1} + \cfrac{1}{a_n}}}}}$$

where a_i are real numbers and $a_i \neq 0$ for $i > 0$. We can write this in a shorter fashion as: $[a_0; a_1, a_2, a_3, \dots, a_{n-1}, a_n]$.

As we said before, we will generate a continued fraction equal to $\sqrt{2}$.

Begin with the identity: $\sqrt{2} + 2 = \sqrt{2} + 2$

Factor the left side and split the 2 on the right side:
$$\sqrt{2}(1 + \sqrt{2}) = 1 + \sqrt{2} + 1$$

Divide both sides by $1 + \sqrt{2}$ to get:
$$\sqrt{2} = 1 + \frac{1}{1 + \sqrt{2}} = [1; 1, \sqrt{2}]$$

Replace $\sqrt{2}$ with $\sqrt{2} = 1 + \dfrac{1}{1 + \sqrt{2}}$ and simplify the terms:
$$\sqrt{2} = 1 + \cfrac{1}{1 + (1 + \cfrac{1}{1 + \sqrt{2}})} = 1 + \cfrac{1}{2 + \cfrac{1}{1 + \sqrt{2}}} = [1; 2, 1, \sqrt{2}]$$

Continue this process. The pattern now becomes clear.

$$\sqrt{2} = 1 + \cfrac{1}{2 + \cfrac{1}{2 + \cfrac{1}{1 + \sqrt{2}}}} = [1; 2, 2, 1, \sqrt{2}\,], \text{ and so on.}$$

Eventually, we conclude with the following:

$$\sqrt{2} = 1 + \cfrac{1}{2 + \cfrac{1}{2 + \cfrac{1}{2 + \ldots}}} = [1; 2, 2, 2, \ldots]$$

Thus we have a periodic continued fraction for $\sqrt{2}$:

$$(\sqrt{2} = [\,1; 2, 2, 2, \ldots] = [1; \overline{2}\,])$$

There are continued fractions equal to some famous numbers such as Euler's e ($e = 2.7182818284590452353 \ldots$)[2] and the famous π ($\pi = 3.1415926535897932384 \ldots$):

$$e = 2 + \cfrac{1}{1 + \cfrac{1}{2 + \cfrac{1}{1 + \cfrac{1}{1 + \cfrac{1}{4 + \cfrac{1}{1 + \cfrac{1}{1 + \cfrac{1}{6 + \ldots}}}}}}}}$$

$$= [2; 1, 2, 1, 1, 4, 1, 1, 6, 1, 1, 8, 1, 1, 10, \ldots] = [2; \overline{1, 2n, 1}]$$

---

2. The number e is the base of the system of natural logarithms. When n increases without limit, then e is the limit of the sequence $\left(1 + \cfrac{1}{n}\right)^n$. The symbol e was introduced by the Swiss mathematician Leonhard Euler (1707–1783) in 1748. In 1761 the German mathematician Johann Heinrich Lambert (1728–1777) showed that e is irrational, and in 1873 the French mathematician Charles Hermite (1822–1901) proved e is a transcendental number.

A transcendental number is a number that is not the root of *any* integer polynomial equation, meaning that it is not an algebraic number of any degree. This definition guarantees that every transcendental number must also be irrational.

Here are two ways that π can be expressed as a continued fraction.[3]

$$\pi = \cfrac{4}{1 + \cfrac{1^2}{2 + \cfrac{3^2}{2 + \cfrac{5^2}{2 + \cfrac{7^2}{2 + \cfrac{9^2}{\ldots}}}}}}$$

$$\frac{\pi}{2} = 1 + \cfrac{1}{1 + \cfrac{1 \cdot 2}{1 + \cfrac{2 \cdot 3}{1 + \cfrac{3 \cdot 4}{1 + \cfrac{4 \cdot 5}{1 + \ldots}}}}}$$

Sometimes we have continued fractions representing these famous numbers that do not seem to have a distinctive pattern:

$$\pi = 3 + \cfrac{1}{7 + \cfrac{1}{15 + \cfrac{1}{1 + \cfrac{1}{292 + \cfrac{1}{1 + \cfrac{1}{1 + \cfrac{1}{1 + \cfrac{1}{2 + \cfrac{1}{1 + \cfrac{1}{3 + \ldots}}}}}}}}}}$$

$\pi = [3; 7, 15, 1, 292, 1, 1, 1, 2, 1, 3, 1, 14, 2, 1, 1, 2, 2, 2, 2, 1, 84, 2, \ldots]$.

We have now set the stage for the golden ratio.[4] Can we express this Fibonacci-related ratio as a continued fraction?

Let's try to use these continued-fraction techniques with ϕ.

---

3 For more on the various representations of π see A. S. Posamentier and I. Lehmann, π: *A Biography of the World's Most Mysterious Number* (Amherst, NY: Prometheus Books, 2004).

4. Remember $\phi = 1.61803\ 39887498948482\ldots$.

Actually, we will use our now well-known relationship for ϕ to produce one of the nicest continued fractions. Begin with this relationship: $\phi = 1 + \dfrac{1}{\phi}$.

We can now replace the ϕ in the denominator on the right side with its equal: $1 + \dfrac{1}{\phi}$.

This gives us: $\phi = 1 + \dfrac{1}{1 + \dfrac{1}{\phi}}$.

Continuing this process will give us the following:

$$\phi = 1 + \frac{1}{\phi} = [\, 1;\, \phi\,]$$

$$\phi = 1 + \cfrac{1}{1 + \cfrac{1}{\phi}} = [\, 1;\, 1, \phi\,]$$

$$\phi = 1 + \cfrac{1}{1 + \cfrac{1}{1 + \cfrac{1}{\phi}}} = [\, 1;\, 1, 1, \phi\,]$$

$$\phi = 1 + \cfrac{1}{1 + \cfrac{1}{1 + \cfrac{1}{1 + \cfrac{1}{\phi}}}} = [\, 1;\, 1, 1, 1, \phi\,]$$

$$\phi = 1 + \cfrac{1}{1 + \cfrac{1}{1 + \cfrac{1}{1 + \cfrac{1}{1 + \cfrac{1}{\phi}}}}} = [\, 1;\, 1, 1, 1, 1, \phi\,]$$

$$\phi = 1 + \cfrac{1}{1 + \cfrac{1}{1 + \cfrac{1}{1 + \cfrac{1}{1 + \cfrac{1}{1 + \cfrac{1}{\phi}}}}}} = [\,1; 1, 1, 1, 1, 1, \phi\,]$$

$$\phi = 1 + \cfrac{1}{1 + \cfrac{1}{1 + \cfrac{1}{1 + \cfrac{1}{1 + \cfrac{1}{1 + \cfrac{1}{\phi}}}}}} = [\,1; 1, 1, 1, 1, 1, 1, \phi\,]$$

$$\phi = 1 + \cfrac{1}{1 + \cfrac{1}{1 + \cfrac{1}{1 + \cfrac{1}{1 + \cfrac{1}{1 + \cfrac{1}{1 + \ldots}}}}}} = [\,1; 1, 1, 1, 1, 1, \ldots\,] = [\,\overline{1}\,]$$

and so on.

Thus we now have a continued fraction equal to $\dfrac{1}{\phi}$ from our value for ϕ:

$$\frac{1}{\phi} = \cfrac{1}{1 + \cfrac{1}{1 + \cfrac{1}{1 + \cfrac{1}{\cdots}}}} = [\,0; 1, 1, 1, 1, 1, 1, \ldots\,] = [\,0; \overline{1}\,]$$

This has to be the nicest continued fraction of all, since the golden section continued fraction ($\phi = \frac{\sqrt{5}+1}{2} \approx 1.6180398874989$) and its reciprocal ($\frac{1}{\phi} = \frac{\sqrt{5}-1}{2} \approx 0.61803398874989$) are comprised of all 1s.

On the other hand, these continued fractions, though elegant, are very slow to approach their true value. You will need many terms to get a good approximation. We will do this with the help of the Fibonacci numbers:

$$\phi_1 = 1 + \frac{1}{1} = 1 + \frac{F_0}{F_1} = \frac{2}{1} = \frac{F_2}{F_1} = 2$$

$$\phi_2 = 1 + \cfrac{1}{1+\cfrac{1}{1}} = 1 + \frac{1}{\phi_1} = 1 + \frac{1}{2} = 1 + \frac{F_1}{F_2} = \frac{3}{2} = \frac{F_3}{F_2} = 1.5$$

$$\phi_3 = 1 + \cfrac{1}{1+\cfrac{1}{1+\cfrac{1}{1}}} = 1 + \frac{1}{\phi_2} = 1 + \frac{2}{3} = 1 + \frac{F_2}{F_3} = \frac{5}{3} = \frac{F_4}{F_3} = 1.\overline{6}$$

$$\phi_4 = 1 + \cfrac{1}{1+\cfrac{1}{1+\cfrac{1}{1+\cfrac{1}{1}}}} = 1 + \frac{1}{\phi_3} = 1 + \frac{3}{5} = 1 + \frac{F_3}{F_4} = \frac{8}{5} = \frac{F_5}{F_4} = 1.6$$

$$\phi_5 = 1 + \cfrac{1}{1+\cfrac{1}{1+\cfrac{1}{1+\cfrac{1}{1+\cfrac{1}{1}}}}} = 1 + \frac{1}{\phi_4} = 1 + \frac{5}{8} = 1 + \frac{F_4}{F_5} = \frac{13}{8} = \frac{F_6}{F_5} = 1.625$$

$$\phi_6 = 1 + \cfrac{1}{1 + \cfrac{1}{1 + \cfrac{1}{1 + \cfrac{1}{1 + \cfrac{1}{1 + \cfrac{1}{1}}}}}} = 1 + \cfrac{1}{\phi_5} = 1 + \cfrac{8}{13} = 1 + \cfrac{F_5}{F_6} = \cfrac{21}{13} = \cfrac{F_7}{F_6}$$

$$= 1.\overline{615384}$$

$$\phi_7 = 1 + \cfrac{1}{1 + \cfrac{1}{1 + \cfrac{1}{1 + \cfrac{1}{1 + \cfrac{1}{1 + \cfrac{1}{1}}}}}} = 1 + \cfrac{1}{\phi_6} = 1 + \cfrac{13}{21} = 1 + \cfrac{F_6}{F_7} = \cfrac{34}{21} = \cfrac{F_8}{F_7}$$

$$= 1.\overline{619047}$$

The *n*th case is then:

$$\phi_n = 1 + \cfrac{1}{n \quad \cdot \quad 1 + \cfrac{1}{1 + \cfrac{1}{1 + \cfrac{1}{1 + \cfrac{1}{1 + \cdots}}}}} = 1 + \cfrac{1}{\phi_{n-1}} = 1 + \cfrac{F_{n-1}}{F_n} = \cfrac{F_{n+1}}{F_n}$$

We then see that the limiting value of the continued fraction for ϕ_n is:[5]

$$\lim_{n \to \infty} \frac{F_{n+1}}{F_n} = \frac{\sqrt{5}+1}{2} = \phi \text{ (See chapter 4.)}$$

---

5. This reads "the limiting value of $\frac{F_{n+1}}{F_n}$, as n gets larger and larger and approaches ∞, equals $\frac{\sqrt{5}+1}{2} = \phi$."

In 1968 Joseph S. Madachy[6] introduced a new constant μ, the continued fraction

$$\mu = 1 + \cfrac{1}{2 + \cfrac{3}{5 + \cfrac{8}{13 + \cfrac{21}{34 + \cfrac{55}{89 + \cfrac{144}{233 + \cfrac{377}{\cdots}}}}}}}$$

$= 1.3941865502287836729028896495777209667374096430683\ldots$

We see that the terms of this continued fraction are successive Fibonacci numbers.

The famous Scottish mathematician Robert Simson (1687– 1768), who took Euclid's *Elements* and wrote an English-language book on it, which is largely responsible for the development of the foundation of the high school geometry course taught in the United States, was the first to recognize that the ratio $\dfrac{F_{n+1}}{F_n}$ of two consecutive Fibonacci numbers will approach the value ϕ of the golden ratio.

Consider the following list of fractions (figure 5-1) to see the value of ϕ being approached.

	(ϕ_0)	ϕ_1	ϕ_2	ϕ_3	ϕ_4	ϕ_5	ϕ_6	ϕ_7	$\cdots$	ϕ
Numerator	(1)	2	3	5	8	13	21	34		
Denominator	(1)	1	2	3	5	8	13	21		
				F_1	F_2	F_3	F_4	F_5	F_6	F_7

Figure 5-1

6. "Recreational Mathematics," *Fibonacci Quarterly* 6, no. 6 (1968): 385–92.

In order to see how the shape of the resulting rectangle gets to look more and more like a golden rectangle, we provide you with a series of drawings. The width *b* remains constant, while the length *a* varies. Each time the rectangle takes on the dimensions of the successive Fibonacci numbers. At the far right, you have the true golden rectangle for comparison. So with the numerator and denominator of ϕ_i (with $i = 1, 2, 3, \ldots, 7$), you can see the development of the golden rectangle (figure 5-2).

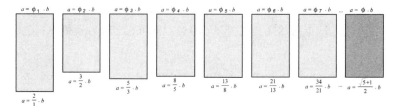

Figure 5-2

You will notice that $\phi_5 = \frac{13}{8} = 1.625$ is a good approximation for $\phi = 1.6180339887498948482 \ldots$. While for $\phi_7 = \frac{34}{21} = 1.\overline{619047}$, we can hardly see the difference between the rectangle formed and the actual golden rectangle. In figure 5-3 the left-side rectangle is the approximation, and the right-side rectangle is the golden one.

Figure 5-3

A Nest of Radicals

Interestingly enough, the 1s play another role in the value of ϕ. Consider the "nest of radicals"

$$\sqrt{1+\sqrt{1+\sqrt{1+\sqrt{1+\cdots}}}}$$

and let's see how we might find the value of it.

This is an analogous method to the one we used to evaluate an infinite continued fraction, since this nest also goes on infinitely long. So we begin by setting this value equal to x :

$$x = \sqrt{1+\sqrt{1+\sqrt{1+\sqrt{1+\cdots}}}}$$

We then square both sides of this equation to get:

$$x^2 = 1+\sqrt{1+\sqrt{1+\sqrt{1+\sqrt{1+\cdots}}}}$$

Since

$$x = \sqrt{1+\sqrt{1+\sqrt{1+\sqrt{1+\cdots}}}}$$

we can replace the nest of radicals in the above equation with x to get $x^2 = 1+x$, which can be written as $x^2 - x - 1 = 0$.

The positive value of x is then $\frac{1+\sqrt{5}}{2} = \phi$. So we then have

$$\phi = \sqrt{1+\sqrt{1+\sqrt{1+\sqrt{1+\cdots}}}}$$

Aha! Another surprising appearance of the golden ratio!

Fibonacci and Lucas Numbers

Let us return now to the Lucas numbers. We would expect to find a relationship between the ratios of consecutive Lucas numbers and the golden ratio, just as we have for the Fibonacci numbers. In figure 5-4, you will notice that both sequences[7] approach the golden ratio.

With that being the case, we ought to be able to use continued fractions with the Lucas numbers also to generate the approach to the golden ratio. Consider the following ratios of consecutive Lucas numbers:

$$\frac{3}{1} = 1 + \frac{2}{1} = 1 + \cfrac{1}{\cfrac{1}{2}}$$

$$\frac{4}{3} = 1 + \frac{1}{3} = 1 + \frac{1}{1+2} = 1 + \cfrac{1}{1+\cfrac{1}{\cfrac{1}{2}}}$$

$$\frac{7}{4} = 1 + \frac{3}{4} = 1 + \cfrac{1}{\cfrac{4}{3}} = 1 + \cfrac{1}{1+\cfrac{1}{1+\cfrac{1}{\cfrac{1}{2}}}}$$

This, in a short version, can be written as $[1; 1, 1, 1, \ldots, \frac{1}{2}]$.

The fact that the last term here is $\frac{1}{2}$, rather than 1 as it is with the Fibonacci numbers, becomes relatively insignificant as the continued fraction increases in terms.

The study of continued fractions holds many other fascinating surprises. But for now we have had the pleasure of using this technique to further explore the Fibonacci and Lucas numbers.

---

7. Rounded to nine decimal places.

$\dfrac{F_{n+1}}{F_n}$	$\dfrac{L_{n+1}}{L_n}$
$\dfrac{1}{1} = 1.000000000$	$\dfrac{3}{1} = 3.000000000$
$\dfrac{2}{1} = 2.000000000$	$\dfrac{4}{3} = 1.333333333$
$\dfrac{3}{2} = 1.500000000$	$\dfrac{7}{4} = 1.750000000$
$\dfrac{5}{3} = 1.666666667$	$\dfrac{11}{7} = 1.571428571$
$\dfrac{8}{5} = 1.600000000$	$\dfrac{18}{11} = 1.636363636$
$\dfrac{13}{8} = 1.625000000$	$\dfrac{29}{18} = 1.611111111$
$\dfrac{21}{13} = 1.615384615$	$\dfrac{47}{29} = 1.620689655$
$\dfrac{34}{21} = 1.619047619$	$\dfrac{76}{47} = 1.617021277$
$\dfrac{55}{34} = 1.617647059$	$\dfrac{123}{76} = 1.618421053$
$\dfrac{89}{55} = 1.6182181618$	$\dfrac{199}{123} = 1.617886179$
$\dfrac{144}{89} = 1.617977528$	$\dfrac{322}{199} = 1.618090452$
$\dfrac{233}{144} = 1.618055556$	$\dfrac{521}{322} = 1.618012422$
$\dfrac{377}{233} = 1.618025751$	$\dfrac{843}{521} = 1.618042226$
$\dfrac{610}{377} = 1.618037135$	$\dfrac{1,364}{843} = 1.618030842$
$\dfrac{987}{610} = 1.618032787$	$\dfrac{2,207}{1,364} = 1.618035191$

Figure 5-4

Chapter 6

A Potpourri of Fibonacci Number Applications

W e will now take a brief tour through some rather unusual and sundry applications and sightings of the ubiquitous Fibonacci numbers. Some will be serious and others lighthearted. These examples are merely a sampling of the wide variety of uses and appearances in which the Fibonacci numbers have been found to manifest themselves. This is by no means meant to be a complete collection. Such a collection would be limitless!

Business Applications

We have seen evidence of the Fibonacci numbers popping up in the most unexpected places. Now, we go searching for them in the volatile world of the stock market.

> All human activities have three distinctive features, pattern, time and ratio, all of which observe the Fibonacci summation series
> —R. N. Elliott

Investor confidence was at an all-time low after the stock market's Great Crash of 1929, whereupon many Americans viewed the stock market as a costly roll of the dice. Nonetheless, a little-

known, albeit successful, accountant and engineer, Ralph Nelson
Elliott (1871–1948), decided to pore over decades of stock per-
formance charts and trace trader movement in an attempt to make
sense of the crash. As his research progressed, distinct and repetitive zigzagging patterns began to emerge as Elliott studied the ups
and downs, or rather, ebb and flow, of market behavior. He dubbed
these patterns "waves" and categorized them as either "impulsive"
or "corrective" waves that could be measured and used to forecast
market behavior.

Figure 6-1
Ralph Nelson Elliott.

In a letter to Charles Collins, the publisher of a national market
newsletter of the day with an extensive following, which included
R. N. Elliott, the enthusiastic Elliott wrote that there was "a much
needed complement to Dow theory" that he called his "wave theory." This correspondence began in November 1934 when Elliott
wrote to Collins about his "discoveries" with the hope that he
would find support from Collins. Collins was impressed by the accuracy of Elliott's analysis and invited him to Detroit to explain
the process in greater detail. Although Elliott's insistence that all
market decisions should be based on wave theory prevented
Collins from directly employing Elliott, he did help him to estab-

lish an office on Wall Street. Later, in 1938, Collins wrote a book-let, under Elliott's name, titled *The Wave Principle*. Elliott went on to further publicize his theory through various letters and magazine articles, including publication in the *Financial World* magazine. Then, in 1946, Elliott expanded on *The Wave Principle* with a book called *Nature's Law—The Secret of the Universe*.

Elliott's famous, influential booklet, *The Wave Principle*, is based on the belief that market behavior should not be seen as ran-dom or chaotic but rather as a natural reflection of investor confi-dence, or lack thereof, (impulsive waves) and self-sustaining mar-ket mechanisms (corrective waves). In layman's terms, Elliott felt the market, like "other things in the universe," moves in predict-able cycles once its patterns of behavior have been established and made visible to the trained eye. At this point, it should come as no surprise that the Fibonacci numbers should be visible within those patterns.

What may be surprising is just how pervasive the Fibonacci numbers are in Elliott's analysis of the stock market. Tracing most bear markets, he found that they move in a series of 2 impulsive waves and 1 corrective wave, for a total of 3 waves (sound famil-iar?). On the other hand, his research showed that a bull market usually jumps upward in 3 impulsive waves and 2 corrective waves, for a total of 5 waves (recognize these numbers?). A com-plete cycle would be the total of these moves, or 8 waves. The re-lationship of the Fibonacci sequence to the wave principle does not end there.

Elliott's wave principle also states that each "major" wave can be further subdivided into "minor" and "intermediate" waves. A regular bear market has 13 intermediate waves. Yes (as you might have suspected), a bull market has 21 intermediate waves, for a total of 34. Continuing on with the sequence, there are 55 minor waves in a bear market and 89 in a bull market, for a total of 144. Elliott himself was not at all surprised to find the Fibonacci se-quence while tracing these market waves because he had always believed that "the stock market is a creation of man and therefore reflects human idiosyncrasy." However, it was only when Elliott began to study the relationships between the size of the waves for his second work, *Nature's Law—The Secret of the Universe*, that he realized his new discoveries were based on "a law of nature known to the designers of the Great Pyramid 'Gizeh,' which may have been constructed 5,000 years ago." (See chapter 7.)

Of course, he was talking about the golden ratio. Elliott believed that this most famous of ratios, along with other Fibonacci-generated ratios, could be used to predict stock prices with astounding accuracy. In order to understand how he reached this conclusion, it is important to investigate what these Fibonacci ratios are. Consider the results when dividing the Fibonacci numbers by their two immediate successors; can you see a pattern emerge after the first several calculations? (See figure 6-2.)

It is clear that after the first several columns, the results in each of the succeeding columns approximate .2360, .3820, and the golden ratio, .6180. Written as percentages, these numbers translate to: 23.6 percent, 38.2 percent, and 61.8 percent and are called Fibonacci percentages. It is interesting to notice a very curious property: the sum of the first two percentages equals the third percentage (.2360 + .3820 = .6180); and the last two percentages (.3820 and .6180) happen to have a sum of 100 percent, a fact that can be proved algebraically.

Elliott's premise, or the Fibonacci indicator, is that the ratio, or the proportional relationship, between two waves could be used to indicate stock prices. Elliott found that an initial wave up in price was usually followed by a second wave downward, or what market analysts call a "retracement" of the initial surge. Upon closer inspection, these retracements seemed, more often than not, to be Fibonacci percentages of the initial price surge, with, 61.8 percent as the highest retracement. Once again, the golden ratio reigns supreme!

n	F_n	$\dfrac{F_n}{F_{n+1}}$	$\dfrac{F_n}{F_{n+2}}$	$\dfrac{F_n}{F_{n+3}}$
1	1	$\dfrac{1}{1} = 1.00000000$	$\dfrac{1}{2} = .500000000$	$\dfrac{1}{3} \approx .33333333$
2	1	$\dfrac{1}{2} = .500000000$	$\dfrac{1}{3} \approx .33333333$	$\dfrac{1}{5} = .20000000$
3	2	$\dfrac{2}{3} \approx .666666667$	$\dfrac{2}{5} = .40000000$	$\dfrac{2}{8} = .25000000$
4	3	$\dfrac{3}{5} = .600000000$	$\dfrac{3}{8} = .37500000$	$\dfrac{3}{13} \approx .230769231$
5	5	$\dfrac{5}{8} = 0.625000000$	$\dfrac{5}{13} \approx .384615385$	$\dfrac{5}{21} \approx .238095238$
6	8	$\dfrac{8}{13} \approx 0.615384615$	$\dfrac{8}{21} \approx .380952381$	$\dfrac{8}{34} \approx .235294118$
7	13	$\dfrac{13}{21} \approx .619047619$	$\dfrac{13}{34} \approx .382352941$	$\dfrac{13}{55} \approx .236363636$
8	21	$\dfrac{21}{34} \approx .617647059$	$\dfrac{21}{55} \approx .381818182$	$\dfrac{21}{89} \approx .235955056$
9	34	$\dfrac{34}{55} \approx .618181818$	$\dfrac{34}{89} \approx .382022471$	$\dfrac{34}{144} \approx .236111111$
10	55	$\dfrac{55}{89} \approx .617977528$	$\dfrac{55}{144} \approx .381944444$	$\dfrac{55}{233} \approx .236051502$
11	89	$\dfrac{89}{144} \approx .618055555$	$\dfrac{89}{233} \approx .381974249$	$\dfrac{89}{377} \approx .236074271$
12	144	$\dfrac{144}{233} \approx .618025751$	$\dfrac{144}{377} \approx .381962865$	$\dfrac{144}{610} \approx .236065574$
13	233	$\dfrac{233}{377} \approx .618037135$	$\dfrac{233}{610} \approx .381967213$	$\dfrac{233}{987} \approx .236068896$

Figure 6-2

While many of today's market analysts incorporate wave patterns and retracements to forecast the market, some analysts take the Fibonacci numbers a step further. They scan previous dates and results looking for the Fibonacci numbers. There might be 34 months or 55 months between a major high or low, or as few as 21 or 13 days between minor fluctuations. A trader might look for repetitive patterns and try to establish a Fibonacci relationship between dates to help forecast the future market.

Before you scoff at the importance of the Fibonacci numbers in the stock market, consider that Elliott, at the age of sixty-seven and without the aid of any computers, forecast the end of a bear market decline from 1933–1935 to the *exact day*.

By the way, while considering the applications of the Fibonacci numbers in the financial world, even one of the main vehicles by which many of these trades are made, the credit card, measures very close to the golden rectangle—55 mm by 86 mm is actually only 3 mm short in the length from fitting a Fibonacci ratio approximation.

A book by Robert Fischer, *Fibonacci Applications and Strategies for Traders*,[1] expounds on the use of the Fibonacci numbers in investment strategies. It is a complete, hands-on resource that shows how to measure price and time signals quickly and with accuracy, by using the logarithmic spiral. He explains how to act on these analyses. This innovative trading guide first takes a fresh look at the classic principles and applications of the Elliott wave theory, then introduces the Fibonacci sequence, exploring its appearance in many of the fields discussed here, and then in stock and commodity trading.

Fischer then explains how the Fibonacci sequence is used to measure equity and commodity price swings and to forecast short- and long-term correction targets. You can learn how to accurately analyze price targets on market extensions, and how, by using the author's logarithmic spirals, to develop price and time analyses with some precision. The book claims to enable you to calculate and predict key turning points in the commodity

1. (New York: Wiley, 1993).

market, analyze business and economic cycles, and discover entry and exit rules that make disciplined trading possible and profitable. We are not suggesting that this is a foolproof method for becoming rich. But the Fibonacci numbers continue to fascinate some in the investment field.

The Vending Machine

A completely different type of application would be setting up a vending machine to dispense a variety of candies. You are told that the cost of each of the candies is to be a multiple of 25 cents. Each type of candy will be a different arrangement (order) of coins deposited, so that the amount of money deposited into the machine and the arrangement of the order in which the coins have been deposited will determine which of the candies has been selected. We need to calculate how many different types of candy the machine can accommodate. Put another way, we need to determine the number of different ways each multiple of 25 cents can be arrived at using the machine's coin slot, which accepts only quarters and half-dollars. For example, for a 75-cent candy selection there are three possible ways to pay: three quarters (QQQ), a quarter and a half-dollar (QH), or a half-dollar and a quarter (HQ). Each of these payments will result in a different type of candy. When we make a chart (figure 6-3) of the various amounts of money that the machine can accept, a curious set of numbers appears. Can you guess what that might be?

Once again, seemingly out of nowhere, the Fibonacci numbers appear in the right-side column, indicating the number of ways by which the various amounts of money can be deposited. This would help the manufacturer of the vending machine predict the number of ways that one could deposit $3.00. The answer would be 12 times 25 cents, or the thirteenth number in the Fibonacci sequence, namely, 233.

Cost of the candies	Number of multiples of 25 cents	List of ways to pay	Number of ways to pay
$.25	1	Q	1
$.50	2	QQ, H	2
$.75	3	QQQ, HQ, QH	3
$1.00	4	QQQQ, HH, QQH, HQQ, QHQ	5
$1.25	5	QQQQQ, HHQ, QHH, HQH, QQQH, HQQQ, QHQQ, QQHQ	8
$1.50	6	QQQQQQ, QQQQH, QQQHQ, QQHQQ, QHQQQ, HQQQQ, QQHH, HHQQ, HQHQ, QHQH, HQQH, QHHQ, HHH	13
$1.75	7	QQQQQQQ, QQQQQH, QQQQHQ, QQQHQQ, QQHQQQ, QHQQQQ, HQQQQQ, QQQHH, QQHHQ, QHHQQ, HHQQQ, QHQHQ, HQHQQ, QQHQH, HQQQH, HQQHQ, QHQQH, HHHQ, HHQH, HQHH, QHHH	21
$2.00	8	QQQQQQQQ, QQQQQQH, QQQQQHQ, QQQQHQQ, QQQHQQQ, QQHQQQQ, QHQQQQQ, HQQQQQQ, QQQQHH, QQQHHQ, QQHHQQ, QHHQQQ, HHQQQQ, HQHQQQ, QHQHQQ, QQHQHQ, QQQQHH, HQQHQQ, QHQQHQ, QQHQQH, HQQQHQ, QHQQQH, HQQQQH, QQHHH, HQQHH, HHQQH, HHHQQ, HHHQQ, QHQHH, HQHQH, HHQHQ, QHHQH, HQHHQ, QHHHQ, HHHH	34

Figure 6-3

Climbing a Staircase

An analogous situation arises with the following problem. Suppose you are about to use a staircase with n stairs and you can either take one step at a time or two steps at a time. The number of different ways of climbing the stairs (C_n) will be a Fibonacci number—much the same way as the previous situation with the vending machine.

Suppose $n = 1$, which means there is only one stair to climb. Then the answer is easy: there is only **one** way to climb the stair.

If $n = 2$, and there are two stairs, then there are **two** different ways to climb the steps: two single steps or one double step.

If $n = 3$, and there are three steps, then there are **three** different ways to climb the steps: 1 step + 1step + 1step, 1 step + 2 steps, or 2 steps + 1 step. This can be generalized (now that you are experts at counting these stair-climbing possibilities). For the staircase of n steps (where $n > 2$), the number of ways of climbing them is C_n. After you take the first step up on a single stair, there are $n - 1$ stairs left to climb. So the number of ways to climb the remainder of the staircase is C_{n-1}. If you begin your climb with a double step (i.e., 2 steps have been climbed), then there are C_{n-2} ways to climb the rest of the staircase. Thus the number of ways to climb the staircase of n steps is the sum of these: $C_n = C_{n-1} + C_{n-2}$. Does this remind you of the Fibonacci recursive relationship? The values of C_n will follow the Fibonacci pattern, with $C_n = F_{n+1}$. You can check this with smaller numbers of stairs and see that it works, or you can compare this to the previous vending machine problem and see that it is the equivalent of that—with the 25-cent units analogous to the stairs.

On the other hand, we can inspect a rather large example. The Empire State Building is 1,453 feet 8 and 9/16th inches tall from street level to its highest point—a lightning rod. There are annual races up the building's 1,860 stairs. In how many ways can an Empire State Building stairs runner reach the topmost step?

It is an unbelievably large number:

$C_n = C_{1,860} = F_{1,861} = 37{,}714{,}947{,}112{,}431{,}814{,}322{,}507{,}744{,}749{,}931{,}$
049,632,797,687,008,623,480,871,351,609,764,568,156,193,373,
680,151,232,412,945,298,517,190,425,833,936,823,942,275,395,
680,820,896,518,732,120,268,852,036,861,867,624,728,128,920,
239,509,015,217,615,431,571,741,968,260,146,431,901,232,750,
464,530,968,296,717,544,866,475,402,917,320,392,352,090,243,
657,224,327,657,131,325,954,780,580,843,850,283,683,054,714,
131,136,328,674,469,916,443,464,802,738,976,662,616,325,164,
306,656,544,521,133,547,290,540,333,738,912,142,760,761
$\approx 3.771494711 \cdot 10^{388}$

Painting a House Creatively

The stories of an n-story house are to be painted blue and yellow, with the stipulation that no two adjacent stories are blue. However, both of them may be yellow. We will let a_n represent the number of possible colorings for a house with n stories ($n \geq 1$).

1 story: 2 stories: 3 stories:
2 possibilities   3 possibilities   5 possibilities

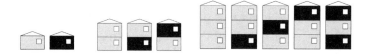

4 stories:
8 possibilities

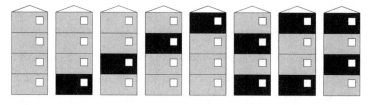

Figure 6-4

In figure 6-4, you will see the possible ways to paint a house of various story heights. Now suppose we have a building with five stories. The fifth story can be painted either yellow or blue. If it gets a yellow color, then the remaining stories can be painted like a four-story house. If it is painted blue, the fourth story must get yellow and the first three stories can be painted like a three-story house. Therefore, we find that $a_5 = a_4 + a_3 = 8 + 5 = 13 = F_6 + F_5 = F_7$. Do you recognize the recursive pattern here? It should be reminiscent of the pattern used to generate the Fibonacci numbers.

Ordered Sum of Ones and Twos

Another nice application of the Fibonacci numbers can be seen from the following challenge:

> In how many ways can a natural number, n, be written as an ordered sum of ones and twos?

Let's inspect the chart in figure 6-5 and count the ways in which we can fulfill this task.

n	1	2	3	4	5	6	7
F_n	1	1	2	3	5	8	13
F_{n+1}	1	2	3	5	8	13	21
ordered sum of ones and twos	1	1+1 2	1+1+1 1+2 2+1	1+1+1+1 1+1+2 1+2+1 2+1+1 2+2	1+1+1+1+1 1+1+1+2 1+1+2+1 1+2+1+1 2+1+1+1 1+2+2 2+1+2 2+2+1	1+1+1+1+1+1 1+1+1+1+2 1+1+1+2+1 1+1+2+1+1 1+2+1+1+1 2+1+1+1+1 1+1+2+2 1+2+2+1 2+2+1+1 2+1+2+1 2+1+1+2 1+2+1+2 2+2+2	1+1+1+1+1+1+1 2+1+1+1+1+1 1+2+1+1+1+1 1+1+2+1+1+1 1+1+1+2+1+1 1+1+1+1+2+1 1+1+1+1+1+2 2+2+1+1+1 2+1+2+1+1 2+1+1+2+1 2+1+1+1+2 1+2+2+1+1 1+2+1+2+1 1+2+1+1+2 1+1+2+2+1 1+1+2+1+2 1+1+1+2+2 2+2+2+1 2+2+1+2 2+1+2+2 1+2+2+2

Figure 6-5

We might conclude from the pattern that seems to be evolving in the chart (figure 6-5) that the natural number n can be written in F_{n+1} different ways as an ordered sum of ones and twos, where F_n is the nth Fibonacci number.

For all $n \geq 2$, the representation of n ends with 1 or 2. The previous terms add up to $n - 1$ or $n - 2$. This yields $F_{n+1} = F_n + F_{n-1}$; $n \geq 2$.

Representing Natural Numbers as the Sum of Fibonacci Numbers

Suppose we take any natural number,[2] say, 27, and try to represent this number as the sum of some Fibonacci numbers. We could do it as $27 = 21 + 5 + 1$, or as $27 = 13 + 8 + 3 + 2 + 1$, or even as $27 = 13 + 8 + 5 + 1$, and with other combinations as well. This will be possible for all natural numbers.[3] You might want to try it for other randomly selected natural numbers. What you should find (with enough examples) is that there will be *only one way* of representing any natural number with *nonconsecutive* Fibonacci numbers.[4] In the above example, the first illustration, $27 = 21 + 5 + 1$, is the only representation of 27 with nonconsecutive Fibonacci numbers. Here are some examples of the unique representations of natural numbers with nonconsecutive Fibonacci numbers:

$1 = F_2$

$2 = F_3$

$3 = F_4$

$4 = F_2 + F_4 = 1 + 3$

$5 = F_5$

$6 = F_2 + F_5 = 1 + 5$

$7 = F_3 + F_5 = 2 + 5$

$8 = F_6$

$9 = F_2 + F_6 = 1 + 8$

$10 = F_3 + F_6 = 2 + 8$

$11 = F_4 + F_6 = 3 + 8$

$12 = F_6 + F_4 + F_2 = 8 + 3 + 1$

Covering a Checkerboard

Strangely enough, the Fibonacci numbers can describe the number of ways to tile a 2 by $(n-1)$ checkerboard with 2 by 1 dominoes; that is, each domino will cover exactly two adjacent squares on the checkerboard. Let's take a look at how that can be done for the first few examples of these.

---

2. A natural number is one of the counting numbers: 1, 2, 3, 4, 5, ...
3. There is a proof of this in appendix B.
4. This was proved by the Belgian amateur mathematician Edouard Zeckendorf (1901–1983), and the theorem is known as the Zeckendorf theorem.

There is only one way to tile a 2×1 board:

Figure 6-6

There are two ways to tile a 2×2 board:

Figure 6-7

There are three ways to tile a 2×3 board:

Figure 6-8

There are five ways to tile a 2×4 board:

Figure 6-9

There are eight ways to tile a 2×5 board:

Figure 6-10

There are thirteen ways to tile a 2 × 6 board:

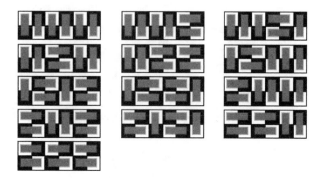

Figure 6-11

There are twenty-one ways to tile a 2 × 7 board:

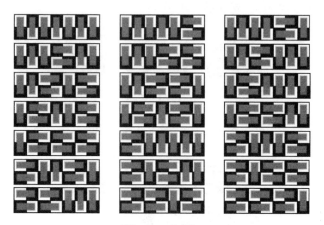

Figure 6-12

While we are on the topic of checkerboards, there is a cute problem with a very useful problem-solving message. Suppose you have an 8 by 8 checkerboard with a pair of opposite-corner squares missing. (See figure 6-13.) Can you show how to cover it with 31 dominoes, each of which is capable of covering exactly two adjacent squares of the checkerboard?

Figure 6-13

As soon as the question is posed, most people try various arrangements of square covering. This may be done with actual tiles or by drawing a graph grid on paper and then shading adjacent squares two at a time. But before long, frustration begins to set in, since this approach cannot be successful.

Here the issue is to go back to the original question. A careful reading of the question reveals that it does not say to do this tile covering; it asks only if it can be done. Yet, because of the way we have been trained, the question is often misread and interpreted as "do it."

A bit of clever insight helps. Ask yourself this question: When a domino tile is placed on the checkerboard, what kind of squares are covered? A black square and a white square must be covered by each domino placed on the checkerboard. Are there an equal number of black and white squares on the truncated checkerboard? No! There are two fewer black squares than white squares. Therefore, it is impossible to cover the truncated checkerboard with the 31 domino tiles, since there must be an equal number of black and white squares. Asking the right questions as you begin to solve a problem and inspecting the problem posed are essential aspects of being successful in problem solving in mathematics.

Fibonacci Numbers and Pythagorean Triangles—Generating Pythagorean Triples

Perhaps the one mathematical relationship that most adults can recall is the famous Pythagorean theorem, which established that the relationship among the sides of a right triangle, whose sides have lengths a, b, and c, is $a^2 + b^2 = c^2$. We may further recall that there are groups of three numbers, such as 3, 4, 5 and 5, 12, 13 as well as 8, 15, 17, that satisfy this relationship, since $3^2 + 4^2 = 9 + 16 = 25 = 5^2$, $5^2 + 12^2 = 25 + 144 = 169 = 13^2$, and $8^2 + 15^2 = 64 + 225 = 289 = 17^2$. Such groups of three numbers that satisfy the Pythagorean theorem are called Pythagorean triples. When the three numbers have no common factor,[5] such as the three examples above, then we call them "primitive Pythagorean triples." An example of a nonprimitive Pythagorean triple is 6, 8, 10, since all the members of this group have a common factor of 2.

From the start, we can state that the Fibonacci numbers have nothing in common with the Pythagorean theorem. They were discovered independently and have no subsequent connection. Yet much to our amazement, we can use the Fibonacci numbers to generate Pythagorean triples. Let's see how this can be done. To make a Pythagorean triple from the Fibonacci numbers, we take any four consecutive numbers in this sequence, such as 3, 5, 8, and 13. We now follow these rules:

1. Multiply the middle two numbers and double the result.
   Here the product of 5 and 8 is 40 then we double this to get **80**. (This is one member of the Pythagorean triple.)
2. Multiply the two outer numbers.
   Here the product of 3 and 13 is **39**. (This is another member of the Pythagorean triple.)
3. Add the squares of the inner two numbers to get the third member of the Pythagorean triple.
   Here $5^2 + 8^2 = 25 + 64 = \underline{\mathbf{89}}$.

---

5. Here and for future references to common factor, we will not consider the number 1, which is always a common factor.

So we have found a Pythagorean triple: 39, 80, 89. We can verify that this is, in fact, a Pythagorean triple by showing that $39^2 + 80^2 = 1,521 + 6,400 = 7,921 = 89^2$. (A proof of this delightful and surprising procedure can be found in appendix B.)

Another Connection between the Fibonacci Numbers and Pythagorean Triples

There is an established method for generating integer sides for a Pythagorean triangle—that is, a right triangle whose sides satisfy the Pythagorean theorem: $a^2 + b^2 = c^2$. This method uses the following set of equations to generate the appropriate values of a, b, and c:

$$a = m^2 - n^2,$$
$$b = 2mn, \text{ and}$$
$$c = m^2 + n^2,$$

where m and n are natural numbers.

For example, if $m = 5$ and $n = 2$, then we can get a Pythagorean triple as follows:

$$a = m^2 - n^2 = 25 - 4 = \underline{21}$$
$$b = 2mn = 2 \cdot 5 \cdot 2 = \underline{20}, \text{ and}$$
$$c = m^2 + n^2 = 25 + 4 = \underline{29},$$

which yields the Pythagorean triple (20, 21, 29).

These three equations, sometimes referred to as *Babylonian formulae*, were already known to have helped the Babylonians generate such Pythagorean triples as (3,456; 3,367; 4,825), and (12,709; 13,500; 18,541).

Remember, a Pythagorean triple (a, b, c) is called *primitive*, if a, b, and c are relatively prime, that is, there is no common factor among a, b, and c. This occurs when m and n are two relatively prime natural numbers, with $m > n$, and of different parity.[6]

6. Two (integer) numbers have the *same parity* if they are both even or both odd.

By now you must be wondering where the Fibonacci numbers come in. Well, suppose we let m and n take on the values of Fibonacci numbers sequentially, beginning with $m = 2$ and $n = 1$. You will find to your amazement that in each case c turns out to also be a Fibonacci number, as you can see in Figure 6-14.

k	m	n	$a = m^2 - n^2$	$b = 2mn$	$c = m^2 + n^2$
1	2	1	3	4	**5**
2	3	2	5	12	**13**
3	5	3	16	30	**34**
4	8	5	39	80	**89**
5	13	8	105	208	**233**
6	21	13	272	546	**610**
7	34	21	715	1,428	**1,597**
8	55	34	1,869	3,740	**4,181**
9	89	55	4,896	9,790	**10,946**
10	144	89	12,815	25,632	**28,657**
11	233	144	33,553	67,104	**75,025**
12	377	233	87,840	175,682	**196,418**
...	...	...	...	...	...

Figure 6-14

We can easily justify this phenomenon. Consider the general case for the two consecutive Fibonacci numbers:

$$m_k = F_{k+2} \text{ and } n_k = F_{k+1}$$

Using the formulas from above, we get the following:

$$a_k = m^2 - n^2 = F_{k+2}^2 - F_{k+1}^2,$$
$$b_k = 2mn = 2F_{k+2} F_{k+1}, \text{ and}$$
$$c_k = m^2 + n^2 = F_{k+2}^2 + F_{k+1}^2$$

which according to relationship 9 in chapter 1 (page 43) is always equal to another Fibonacci number, F_{2k+3}.

Can a Triangle Have Sides with Lengths Represented by Fibonacci Numbers?

Clearly, you can have an equilateral triangle each of whose sides have length 5, a Fibonacci number, or you can have an isosceles triangle with sides 13, 13, and 5—also made up of Fibonacci lengths. However, if the numbers we select to represent the sides of a triangle are all *different* Fibonacci numbers, then the result is quite different. No such triangle can exist! We can show that a triangle can never have all three sides with lengths represented by three distinct Fibonacci numbers.

This can be easily established[7] by invoking the triangle inequality—namely, for a triangle to exist, the sum of any two sides must be greater than the third side: hence, for a triangle with side lengths a, b, and c, the following relationships would be true:

$$a + b > c, b + c > a, \text{ and } c + a > b$$

For any three consecutive Fibonacci numbers, $F_n + F_{n+1} = F_{n+2}$, and so there can be no triangle with sides having measures F_n, F_{n+1}, and F_{n+2}.

In general, consider Fibonacci numbers, F_r, F_s, and F_t, where $F_r \leq F_{s-1}$ and $F_{s+1} \leq F_t$. Since $F_{s-1} + F_s = F_{s+1}$ and $F_r \leq F_{s-1}$, we have $F_r + F_s \leq F_{s+1}$, and since $F_{s+1} \leq F_t$, we have $F_r + F_s \leq F_t$. Therefore, there can be no triangle with sides having measure F_r, F_s, and F_t.

A Multiplication Algorithm Using the Fibonacci Numbers

It is said that the Russian peasants used a rather strange, perhaps even primitive, method to multiply two numbers.[8] It is actually quite simple, yet somewhat cumbersome. Let's take a look at it.

---

7. V. E. Hogatt Jr., *Fibonacci and Lucas Numbers* (Boston: Houghton Mifflin, 1969), p. 85.
8. There is also evidence that this algorithm was used by the ancient Egyptians.

We shall begin with a simple case. If we take the product of two integers, where one of the factors is a power of 2, then the process becomes trivial as you will see from the example of $65 \cdot 32$.[9] We can see that the product of two integers remains constant if one factor is doubled and the other factor is halved. We do this in the following table:

	65	32
$65 \cdot 32 =$	130	16
	260	8
	520	4
	1,040	2
	2,080	1

$65 \cdot 32 = $ **2,080**

Figure 6-15

So products in which a factor is a power of 2 are particularly simple to calculate, since we can read off figure 6-15 that $1 \cdot 2,080 = 2,080$. Thus we have found the product of 65 and 32.

We shall now consider the problem of finding the product of $43 \cdot 92$. This time we are not using a factor that is a power of 2.

Let's work this multiplication together. We begin by setting up a chart of two columns with the two members of the product in the first row (figure 6-16). We place the numbers 43 and 92 at the head of the columns. One column will be formed by doubling each number to get the next, while the other column will take half the number and drop the remainder. For convenience, our first column (the left-side column) will be the doubling column, and the second column will be the halving column. Notice that by halving an odd number such as 23 (the third number in the right-side column) we get 11 with a remainder of 1 and we simply drop the 1. The rest of this halving process should now be clear.

---

9. Remember 32 is a power of 2: $2^5 = 32$.

43	92
86	46
172	**23**
344	**11**
688	**5**
1,376	2
2,752	**1**

Figure 6-16

Find the odd numbers in the halving column (the right column), then get the sum of the partner numbers in the doubling column (the left column). These are highlighted in bold type. This
sum gives you the originally required product of 43 and 92. In
other words, with the Russian peasant's method, we get

$$43 \cdot 92 = 172 + 344 + 688 + 2{,}752 = 3{,}956$$

In the example above, we chose to have the first column be the
doubling column and the second column be the halving column.
We could also have done this Russian peasant's method by halving
the numbers in the first column and doubling those in the second.
See figure 6-17.

43	92
21	**184**
10	368
5	**736**
2	1,472
1	**2,944**

Figure 6-17

To complete the multiplication, we find the odd numbers in the
halving column (in bold type), and then get the sum of their partner
numbers in the second column (now the doubling column). This
gives us $43 \cdot 92 = 92 + 184 + 736 + 2{,}944 = 3{,}956$.

Obviously, you are not expected to do your multiplication in
this high-tech era by copying the Russian peasant's method. However, it should be fun to observe how this primitive system of

arithmetic actually does work. Explorations of this kind are not only instructive but also entertaining.

Here you see what was done in the above multiplication algorithm.[10]

$43 \cdot 92 \quad = (21 \cdot 2 + 1) \cdot 92 = 21 \cdot 2 \cdot 92 + 1 \cdot 92 = 21 \cdot 184 + \quad \mathbf{92} = 3,956$

$21 \cdot 184 = (10 \cdot 2 + 1) \cdot 184 = 10 \cdot 2 \cdot 184 + 1 \cdot 184 = 10 \cdot 368 + \quad \mathbf{184} = 3,864$

$10 \cdot 368 = (5 \cdot 2 + 0) \cdot 368 = 5 \cdot 2 \cdot 368 + 0 \cdot 368 = 5 \cdot 736 + \quad \mathbf{0} = 3,680$

$5 \cdot 736 \quad = (2 \cdot 2 + 1) \cdot 736 = 2 \cdot 2 \cdot 736 + 1 \cdot 736 = 2 \cdot 1,472 + \quad \mathbf{736} = 3,680$

$2 \cdot 1,472 = (1 \cdot 2 + 0) \cdot 1,472 = 1 \cdot 2 \cdot 1,472 + 0 \cdot 1472 = 1 \cdot 2,944 + \quad \mathbf{0} = 2,944$

$1 \cdot 2,944 = (0 \cdot 2 + 1) \cdot 2,944 = 0 \cdot 2 \cdot 2,944 + 1 \cdot 2944 = 0 + \quad \underline{\mathbf{2,944}} = 2,944$

$$\mathbf{3,956}$$

For those familiar with the binary system (i.e., base 2), one can also explain this Russian peasant's method with the following representation.

$$43 \cdot 92 = (1 \cdot 2^5 + 0 \cdot 2^4 + 1 \cdot 2^3 + 0 \cdot 2^2 + 1 \cdot 2^1 + 1 \cdot 2^0) \cdot 92$$
$$= 2^0 \cdot 92 + 2^1 \cdot 92 + 2^3 \cdot 92 + 2^5 \cdot 92$$
$$= 92 + 184 + 736 + 2,944$$
$$= 3,956$$

Whether or not you have a complete understanding of the discussion of the Russian peasant's method of multiplication, you should at least now have a deeper appreciation for the multiplication algorithm you learned in school, even though most people today multiply with a calculator. There are many other multiplication algorithms, yet the one shown here is perhaps one of the strangest. It is through this strangeness that we can appreciate the powerful consistency of mathematics and that allows us to conjure up such an algorithm. Now let us bring in the Fibonacci numbers. With the aid of these numbers, we can develop an analogous multiplication algorithm. The Fibonacci multiplication algorithm will use addi-

---

10. We have defined an algorithm as a step-by-step problem-solving procedure, especially an established, recursive computational procedure for solving a problem in a finite number of steps.

tion instead of doubling. Let us use the same two numbers that we used for the multiplication example above: $43 \cdot 92$.

This time we will place the 92 at the top of the right-side column (figure 6-18). On the left side we will begin with the second 1 of the Fibonacci sequence and list the members of the sequence until we reach the Fibonacci number that is just less than the other number to be multiplied—in this case, 43. So we list the Fibonacci numbers to 34, since the next Fibonacci number, 55, would exceed 43.

In the right-side column, we double 92 to get the next number, and then add every pair of numbers to get the succeeding one (just as we do with the Fibonacci numbers). We now select the numbers in the left-side column that will give us a sum of 43 (the original multiplier). We have selected $1 + 8 + 34 = 43$. The sum of their partners in the right-side column is our desired product:

$$92 + 736 + 3{,}128 = 3{,}956$$

1	92	
2	184	$= 92 + 92$
3	276	$= 184 + 92$
5	460	$= 276 + 184$
8	**736**	$= 460 + 276$
13	1,196	$= 736 + 460$
21	1,932	$= 1{,}196 + 736$
34	**3,128**	$= 1{,}932 + 1{,}196$

Figure 6-18

This may not be the algorithm of choice, but it is fascinating to observe how our Fibonacci numbers can even enable us to do multiplication.

Using the Fibonacci Numbers to Convert Kilometers to, and from, Miles

Most of the world uses kilometers to measure distance, while the United States still holds on to the mile. This requires a conversion of units when one travels in a country in which the measure of distance is not the one we are accustomed to. Such conversions can be done with specially designed calculators or by some "trick" method. This is where the Fibonacci numbers come in. Before we discuss the conversion process between these two units of measure, let's look at their origin.

The mile derives its name from the Latin for thousand: *mille*. It represented the distance that a Roman legion could march in 1,000 paces (which is 2,000 steps). One of these paces was about 5 feet, so the Roman mile was about 5,000 feet. The Romans marked off these miles with stones along the many roads they built in Europe—hence the name, "milestones"! The name *statute mile* goes back to Queen Elizabeth I of England, who redefined the mile from 5,000 feet to 8 furlongs[11] (5,280 feet) by statute in 1593.

The metric system dates back to 1790 when the French National Assembly (during the French Revolution) requested that the French Academy of Sciences establish a standard of measure based on the decimal system, which they did. The unit of length they called a "meter," derived from the Greek word, *metron*, which means measure. Its length was determined to be one ten-millionth of the distance from the North Pole to the equator along a meridian going near Dunkirk, France, and Barcelona, Spain.[12]   Clearly, the

---

11. A furlong is a measure of distance within Imperial units and US customary units. Although its definition has varied historically, in modern terms it equals 660 feet, and is therefore equal to 201.168 meters. There are eight furlongs in a mile. The name furlong derives from the Old English words *furh* (furrow) and *lang* (long). It originally referred to the length of the furrow in one acre of a ploughed open field (a medieval communal field which was divided into strips). The term is used today for distances horses run at a racetrack.
12. There are three fascinating books on this subject: Dava Sobel, *Longitude* (New York: Walker & Co., 1995); Umberto Eco, *The Island of the Day Before* (New York: Harcourt Brace, 1994); and Thomas Pynchon, *Mason & Dixon* (New York: Henry Holt, 1997).

metric system is better suited for scientific use than is the American system of measure. By an act of Congress in 1866 it became "lawful throughout the United States of America to employ the weights and measures of the metric system in all contracts, dealings, or court proceedings" (though it has not been used that often). And there is no such law establishing the use of our mile system.

Now to convert miles to and from kilometers, we need to see how one mile relates to the kilometer. The statute mile (our usual measure of distance in the United States today) is exactly 1,609.344 meters long. This translated into kilometers is 1.609344 kilometers. One the other hand, one kilometer is .621371192 mile long. The nature of these two numbers (reciprocals that differ by almost 1) might remind us of the golden ratio, which is approximately 1.618 and has a reciprocal of approximately .618. Remember, it is the only number whose reciprocal differs from it by exactly 1. This would tell us that the Fibonacci numbers, the ratio of whose consecutive members approaches the golden ratio, might come into play here.

Let's see what the equivalent of five miles would be in kilometers:

$$5 \cdot 1.609344 = 8.04672 \approx 8$$

We could also check to see what the equivalent of eight kilometers would be in miles:

$$8 \cdot .621371192 = 4.970969536 \approx 5$$

This allows us to conclude that approximately five miles is equal to eight kilometers. Here we have two of our Fibonacci numbers.

The ratio of a Fibonacci number to the one before it is approximately ϕ. Therefore, since the relationship between miles and kilometers is very close to the golden ratio, they appear to be almost in the relationship of consecutive Fibonacci numbers. Using this relationship, we would be able to approximately **convert thirteen kilometers to miles**. Replace 13 by the previous Fibonacci number, 8, and thirteen kilometers is about eight miles. Similarly, five kilometers is about three miles and two kilometers is about one mile. The higher Fibonacci numbers will give us a more accurate estimate, since the ratio of these larger consecutive numbers gets closer to ϕ.

Now suppose you want to convert twenty kilometers to miles. We have selected 20, since it is *not* a Fibonacci number. We can express 20 as a **sum of Fibonacci numbers**,[13] convert each number separately, and then add them. Thus, 20 kilometers = 13 kilometers + 5 kilometers + 2 kilometers, which, by replacing 13 by 8, 5 by 3, and 2 by 1, is approximately equal to twelve miles.

To use this process to achieve the reverse, that is, to **convert miles to kilometers**, we write the number of miles as a sum of Fibonacci numbers and then replace each by the next *larger* Fibonacci number. Let's convert twenty miles to kilometers. Thus, 20 miles = 13 miles + 5 miles + 2 miles. Now, replacing each of the Fibonacci numbers with the next larger in the sequence, we get 20 miles ≈ 21 kilometers + 8 kilometers + 3 kilometers = 32 kilometers.

---

13. **Representing Numbers as Sums of Fibonacci Numbers:** It is not a trivial matter for us to conclude that every natural number can be expressed as the sum of other Fibonacci numbers without repeating any one of them in the sum. Let's take the first few Fibonacci numbers to demonstrate this property:

n	The Sum of Fibonacci Numbers equal to n
1	1
2	2
3	3
4	1 + 3
5	5
6	1 + 5
7	2 + 5
8	8
9	1 + 8
10	2 + 8
11	3 + 8
12	1 + 3 + 8
13	13
14	1 + 13
15	2 + 13
16	3 + 13

You should begin to see patterns and also note that we used the fewest number of Fibonacci numbers in each sum in the table above. For example, we could also have represented 13 as the sum of 2 + 3 + 8, or as 5 + 8. Try to express larger natural numbers as the sum of Fibonacci numbers. Each time ask yourself if you have used the fewest number of Fibonacci numbers in the sum. It will be fun to see the patterns that develop. By the way, Edouard Zeckendorf (1901–1983) proved that each natural number is a (unique) sum of nonconsecutive Fibonacci numbers.

There is no need to use the Fibonacci representation of a number that uses the fewest Fibonacci numbers. You can use any combination of numbers whose sum is the number you are converting. For instance, forty kilometers is two times twenty kilometers, and we have just seen that twenty kilometers is twelve miles. So forty kilometers is two times twelve miles, or twenty-four miles (approximately).

Fibonacci Numbers in Physics

The field of optics provides us with a nice application of the Fibonacci numbers.[14] Suppose you place two glass plates face to face (figures 6-19 and 6-20) and wish to count the number of possible reflections. This might be achieved by intermittently covering a surface, say, at the back side. For convenience, we shall number the various surfaces of reflection as in figure 6-20.

Figure 6-19   **Figure 6-20**

The first case, where there are no reflections, is where the light source passes right through both plates of glass (figure 6-21). There is **one** path the light ray takes.

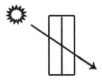

Figure 6-21

14. This problem was posed by L. Moser and M Wyman, "Problem B-6," *Fibonacci Quarterly* 1, no. 1 (February 1963): 74, and solved by L. Moser and J. L. Brown, "Some Reflections," *Fibonacci Quarterly* 1, no. 1 (December 1963): 75.

The next case is where there is one reflection. There can be **two** possible paths that the light can take in this case. It can reflect off the two surfaces shown in figure 6-22.

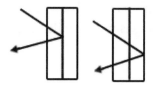

Figure 6-22

Suppose, now, that the ray of light is reflected twice within this set of two plates of glass. Then there will be **three** possible reflections (figure 6-23).

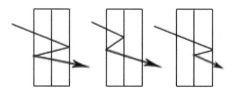

Figure 6-23

If the light ray is reflected three times, then there will be **five** possible paths that the ray can take—as seen in figure 6-24.

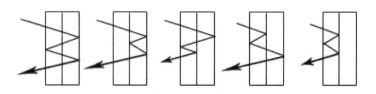

Figure 6-24

By now you will surely anticipate that the next case, that of the light ray taking four reflections, will result in **eight** possible paths—as seen in figure 6-25.

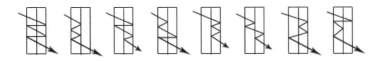

Figure 6-25

You may be tempted to generalize this situation to the Fibonacci numbers. And, if you do, you are, in fact, right! However, you may want to be able to generalize this with a mathematical argument. Let's try to justify this in a mathematical fashion.

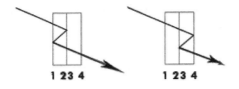

Figure 6-26

Suppose the last reflection occurs at face 1 or 3, as it did with the even number of reflections. The previous reflection had to be off faces 2 or 4. (figure 6-26).

If the last reflection, of n reflections, was off face 1, then there were $n - 1$ reflections prior to this one, or $n - 1$ possible paths (see figure 6-26). If the last reflection (of n reflections) was off face 3, then the previous reflection—the $(n - 1)$st—must have occurred off face 4, and must have had $n - 2$ reflections prior to this one. And so there were $n - 2$ possible paths. Therefore, since both reflections—off face 1 or 3—can be considered the last reflection, we get the number of paths as $(n - 1) + (n - 2)$, which would indicate that F_n is the number of paths. This began with zero reflections giving us **1** path, then one reflection, which gave us **2** paths, and three reflections, which gave us **3** paths. This implies that F_{n+2} indicates the number of paths for n reflections. Again the Fibonacci numbers appear!

Last-Digit Patterns

We mentioned in chapter 1 that there is a pattern among the final digits of the Fibonacci numbers. Namely, they have a period of 60. That is, the first 60 Fibonacci numbers, $F_1 - F_{60}$, have the same last digits (or sometimes called terminal digits), respectively, as the second 60 Fibonacci numbers, $F_{61} - F_{120}$, the third 60 Fibonacci numbers, $F_{121} - F_{180}$, the fourth 60 Fibonacci numbers, $F_{181} - F_{240}$, and so on for each group of 60 Fibonacci numbers. (You can check this with the list of the first 500 Fibonacci numbers provided for you in appendix A.)

Suppose we now consider the last *two* digits of each of the Fibonacci numbers (which are below in bold).

00, **01**, **01**, **02**, **03**, **05**, **08**, **13**, **21**, **34**, **55**, **89**, 1**44**, 2**33**, 3**77**, 6**10**, 9**87**, 1,5**97**, 2,5**84**, 4,1**81**, 6,7**65**, 10,9**46**, 17,7**11**, 28,6**57**, 46,3**68**, 75,0**25**, 121,3**93**, . . .

If we look at the 300th Fibonacci number in appendix A, we find the last two digits repeating beginning with the 301st Fibonacci number. Hence, this period is 300 numbers long. Yes, if we were to look at F_{600}, we would find the pattern continuing. You might be speculating about the period of repetition of the last three digits of the Fibonacci numbers. Well, we won't make you struggle with this computation. For the next few groups of terminal digits, we can conclude the following:

- The period of the last *three* digits of the Fibonacci numbers repeating is 1,500.
- The period of the last *four* digits of the Fibonacci numbers repeating is 15,000.
- The period of the last *five* digits of the Fibonacci numbers repeating is 150,000.

If you do have the time, you can verify this. Or else you can take our word for it.

The First Digits of Fibonacci Numbers

Now that we have inspected the relationship among the terminal digits of the Fibonacci numbers, let's turn to the initial digits. One would assume that there would be an even distribution of the nine numerals among the initial digits of the Fibonacci numbers. If you take the time to inspect the first five hundred Fibonacci numbers in appendix A, you will find that the most frequent numeral is 1, and the least frequent is 9. Figure 6-27 provides a summary of the frequency of the initial digits of the first one hundred Fibonacci numbers.

Initial digit	Frequency (Percent)
1	30
2	18
3	13
4	9
5	8
6	6
7	5
8	7
9	4

Figure 6-27

Quite surprising! The reason lies in the fact that Fibonacci numbers (as you will see in chapter 9 when we investigate the Binet formula) are expressible as powers of $\sqrt{5}$. This, in part, also accounts for the uneven distribution of the Fibonacci numbers throughout the list of natural numbers. What does this actually mean? Within the first one hundred natural numbers (i.e., the numbers from 1 to 100), there are eleven Fibonacci numbers. Amazingly enough, among the next hundred natural numbers (from 101 to 200) there is only one Fibonacci number (144). As we look further down the list of natural numbers, we find that among the next three hundred natural numbers there are only two Fibonacci numbers, and further there are only two Fibonacci numbers between 500 and 1,000 (i.e., 610 and 987). This paucity continues, and you can see it clearly in appendix A. Who would have expected this strange distribution! This is even made more dramatic when seen through a graph. The visualization shows how fast (and steeply) the values grow. The eighteenth Fibonacci number ($F_{18} = 2,584$) still can be read on the graph in figure 6-28 the nineteenth Fibonacci number ($F_{19} = 6,765$) is far off the scale shown in figure 6-28.

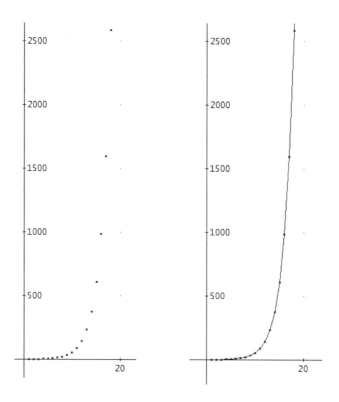

Figure 6-28

A Fibonacci Curiosity of Digits[15]

Here is a strange "coincidence" (or is it one?) that seems to occur for some of the Fibonacci numbers. And it is that the sum of the digits of these Fibonacci numbers, F_n, is equal to n. The table in figure 6-29 lists the Fibonacci numbers where this relationship holds true.

n	F_n	Sum of the Digits
0	0	0
1	1	1
5	5	5
10	55	10
31	1,346,269	31
35	9,227,465	35
62	4,052,739,537,881	62
72	498,454,011,879,264	72

Figure 6-29

This is just a curiosity to further entice you to appreciate these intriguing numbers.

Examining Relationships between Successive Fibonacci Numbers

While we are examining the characteristics of the members of the Fibonacci sequence, we might look at the relationship between two

---

15. Leon Bankoff, "A Fibonacci Curiosity," *Fibonacci Quarterly* 14, no. 1 (1976): 17.

consecutive Fibonacci numbers. Of course, by this point, you know that the ratio of two consecutive Fibonacci numbers gives us a good approximation of the golden ratio. However, there are other relationships worth noting. A quick look at appendix A will verify that no two consecutive Fibonacci numbers are even. Another way of saying this is that they do not have a common factor of 2. As a matter of fact, you will notice that every third Fibonacci number is even, or that there are two odd Fibonacci numbers between every two consecutive even Fibonacci numbers.

Looking at the Fibonacci numbers that have a factor of 3, you will find they are $F_4, F_8, F_{12}, F_{16}, \ldots$. Every fourth Fibonacci number has a factor of 3. Although we already noted this in chapter 1, we recall that no two consecutive Fibonacci numbers have a common factor. (We call two numbers that have no common factor "relatively prime numbers.") We can do a simple proof of this fact here.

- If x and y have a common factor, then it must also be a factor of $x + y$.
- If x and y have a common factor, then it is also a factor of $y - x$.
- If x and y have *no* common factor, then y and $x + y$ also have no common factor. This is because if y and $x + y$ had a common factor, then their *difference,* $y - (x + y)$, which is just equal to $-x$, would also have a common factor, but that is absurd!
- Therefore, if the first two members of a sequence are relatively prime, then all pairs of consecutive members of the sequence are relatively prime.
- We know that F_1 and F_2 are relatively prime (since they have no common factor).
- Therefore, all the members of the sequence F_n and F_{n+1} are relatively prime.

This little proof is a fine example of how a mathematical truth is established.

A Cute Curiosity!

Consider the four consecutive Fibonacci numbers F_7, F_8, F_9, and F_{10}, or specifically, 13, 21, 34, and 55. Now let's look at these numbers in another form. We will break each one down into its prime factors: (13), (7, 3), (2, 17), and (5, 11). If we rearrange this list of factors in ascending order, we get: 2, 3, 5, 7, 11, 13, and 17, which are the first seven prime numbers. The product of these numbers is 510, which just happens to be the Library Dewey Decimal classification number[16] for mathematics! This is a coincidental fun fact!

Special Fibonacci Numbers

Those numbers representing dots that can be arranged to form an equilateral triangle are called triangular numbers. As shown below, the first few triangular numbers are 1, 3, 6, 10, 15, 21, and 28.

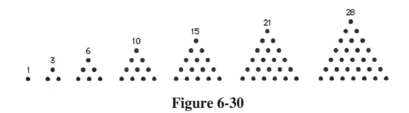

Figure 6-30

A triangular number is one that is an integer of the form $\frac{n(n+1)}{2}$, where $n = 1, 2, 3, \ldots$.

---

16. American librarian Melvil Dewey, born Melville Louis Kossuth Dewey, in Adams Center, New York, on December 10, 1851, is best known as the inventor of what came to be called the "Dewey Decimal System of Classification," which is used in most local and school libraries to catalog books. Devised in 1876 as a system for small libraries, it has the advantage of a limited number of general categories and short call numbers. The system is based on ten classes of subject (000–999), which are then further subdivided. Dewey also promoted the use of the metric system, and helped found the American Library Association in 1876. When Dewey created Columbia University's School of Library Economy in 1887, he began the field of library science in the United States.

However, there are only four Fibonacci numbers that are triangular numbers. They are 1, 3, 21, and 55.

There are only three Lucas numbers that are triangular numbers. They are 1, 3, and 5,778. So even in the realm of triangular numbers, the Fibonacci numbers make an appearance.

The thirty-sixth triangular number[17] is 666, a number that has captured the popular imagination for a long time. (Many associate it with the devil.) The sum of the squares of the first seven prime numbers is 666. That is, $2^2 + 3^2 + 5^2 + 7^2 + 11^2 + 13^2 + 17^2 = 666$.

This unusual number also offers some nice number relationships, such as:

$$666 = 1^6 - 2^6 + 3^6$$
$$666 = 6 + 6 + 6 + 6^3 + 6^3 + 6^3$$
$$666 = 2 \cdot 3 \cdot 3 \cdot 37, \text{ and } 6 + 6 + 6 = 2 + 3 + 3 + 3 + 7$$

And just for the sake of entertainment on the number 666, consider the first six Roman numerals written in order: DCLXVI, which just happens to equal 666.

Of course, this number, 666, is also related to the Fibonacci numbers.

$$F_1 - F_9 + F_{11} + F_{15} = 1 - 34 + 89 + 610 = 666$$

And the analog for the subscripts:

$$1 - 9 + 11 + 15 = 6 + 6 + 6$$

Similarly for the cubes of the Fibonacci numbers:

$$F_1^3 + F_2^3 + F_4^3 + F_5^3 + F_6^3 = 1 + 1 + 27 + 125 + 512 = 666$$

And the analog for the subscripts:

$$1 + 2 + 4 + 5 + 6 = 6 + 6 + 6$$

---

17. The thirty-sixth triangular number $= \dfrac{(36)(36+1)}{2} = 18 \cdot 37 = 666$.

Now let's have this curious number 666 connect to the Fibonacci numbers with the golden ratio—as is fitting to do in this book:

$$-2\sin(666°) = 1.618033988749894848204586834365638117720309179805 7\ldots$$

a very close approximation of the golden ratio, which is

$$\phi = 1.6180339887498948482045868343656381177203091798057\ldots$$

Finally, in 1994 Steve C. Wang published the following mysterious equation in the *Journal of Recreational Mathematics*[18]—with the title "The Sign of the Devil":

$$\phi = -[\sin(666°) + \cos(6 \cdot 6 \cdot 6\,°)]$$

While earlier we had a fine approximation of the golden ratio, this equation, then, places the number 666 in an unquestionable connection with the golden ratio!

The Curiosity of the Eleventh Fibonacci Number

The eleventh Fibonacci number, F_{11}, is 89. This number has many unusual characteristics. For example, if you take any number and find the sum of the squares of the digits, and continuously repeat this process, you will either end up with 1 or 89.[19] We will show how this works in a few cases:

---

18. 26, no. 3 (1994): 201–205.
19. If you take this further you will be led to the number 4, but this is via the number 89. This would continue as:
$$8^2 + 9^2 = 145$$
$$1^2 + 4^2 + 5^2 = 42$$
$$4^2 + 2^2 = 20$$
$$2^2 + 0^2 = 4,$$
which then takes you into a loop back to 89:
$$4^2 = 16$$
$$1^2 + 6^2 = 37$$
$$3^2 + 7^2 = 58$$
$$5^2 + 8^2 = 89$$

For the number 23:

$$2^2 + 3^2 = 13$$
$$1^2 + 3^2 = 10$$
$$1^2 + 0 = 1$$

For the number 54:

$$5^2 + 4^2 = 41$$
$$4^2 + 1^2 = 17$$
$$1^2 + 7^2 = 50$$
$$5^2 + 0 = 25$$
$$2^2 + 5^2 = 29$$
$$2^2 + 9^2 = 85$$
$$8^2 + 5^2 = 89$$

For the number 64:

$$6^2 + 4^2 = 52$$
$$5^2 + 2^2 = 29$$
$$2^2 + 9^2 = 85$$
$$8^2 + 5^2 = 89$$

To convince yourself of this peculiarity, try this with some other numbers. If you are still piqued by this Fibonacci peculiarity, consider the reciprocal of 89:

$\dfrac{1}{89}$ = 0.01123595505617977528089887640449438202247191_01123595505

61797752808988764044943820224719_01123595505617977528089 887640449438202247191_01123595505617977528089887640449438 202247191 ...

$\overline{\rule{0pt}{1em}\hspace{8em}}$

= 0.01123595505617977528089887640449438202247191

You will notice it repeats after forty-four decimal places, or as we say, it has a period of 44. Curiously, at the very middle of this period of forty-four digits you will notice there is the number 89.

$$\underline{0.0112359550561797752808988764044943820224719\underline{1}}$$

Its unusual Fibonacci property is even more surprising. Notice the first few digits of the period, 0, 1, 1, 2, 3, and 5, are the first few Fibonacci numbers. So what happens after that? To see this continue, consider the following:

$$\frac{1}{89} = \frac{0}{10^1} + \frac{1}{10^2} + \frac{1}{10^3} + \frac{2}{10^4} + \frac{3}{10^5} + \frac{5}{10^6} + \frac{8}{10^7} + \frac{13}{10^8} + \cdots$$

$$\frac{1}{89} = \frac{F_0}{10^1} + \frac{F_1}{10^2} + \frac{F_2}{10^3} + \frac{F_3}{10^4} + \frac{F_4}{10^5} + \frac{F_5}{10^6} + \frac{F_6}{10^7} + \frac{F_7}{10^8} + \cdots$$

This may be seen, perhaps, more concretely when written as decimals:

```
        0
    + .01
    + .001
    + .0002
    + .00003
    + .000005
    + .0000008
    + .00000013
    + .000000021
    + .0000000034
    + .00000000055
    + .000000000089
    + .0000000000144
    + .00000000000233
    + .000000000000377
    + .0000000000000610
    + .0000000000000987
```

.01123595505617977 . . .

This will continue to generate the above period of forty-four digits if we take the decimal approximations further.

But the question might be asked: Why does this happen? We can do some numerical manipulations that will show how the

Fibonacci numbers tie in with this remarkable number 89—itself the eleventh Fibonacci number.

We begin with the following equality:

$$10^2 = 89 + 10 + 1 \tag{I}$$

We will multiply both sides of (I) by 10:

$$10^3 = 89 \cdot 10 \ + 10^2 + 10$$

and then substitute (I) in the resulting equality to get:

$$10^3 = 89 \cdot 10 \ + \ 89 + 10 + 1 + 10$$
$$10^3 = 89 \cdot 10 \ + 89 + \ 2 \cdot 10 \ + 1 \tag{II}$$

Continuing this process, we will multiply both sides of (II) by 10:

$$10^4 = 89 \ \cdot \ 10^2 + 89 \ \cdot \ 10 + 2 \ \cdot \ 10^2 + 1 \ \cdot \ 10$$
$$10^4 = 89 \ \cdot \ 10^2 + 89 \ \cdot \ 10 + 2(89 + 10 + 1) + 1 \ \cdot \ 10$$

and then substitute (I) again, where appropriate to get:

$$10^4 = 89 \ \cdot \ 10^2 + 89 \ \cdot \ 10 + 89 \ \cdot \ 2 + 3 \ \cdot \ 10 + 2 \tag{III}$$

Continuing this process gives us the following:

$$10^5 = 89 \ \cdot \ 10^3 + 89 \ \cdot \ 10^2 + 89 \ \cdot \ 2 \ \cdot \ 10 + 3 \ \cdot \ 10^2 + 2 \ \cdot \ 10$$
$$10^5 = 89 \ \cdot \ 10^3 + 89 \ \cdot \ 10^2 + 89 \ \cdot \ 2 \ \cdot \ 10 + 3(89 + 10 + 1) + 2 \ \cdot \ 10$$
$$10^5 = 89 \ \cdot \ 10^3 + 89 \ \cdot \ 10^2 + 89 \ \cdot \ 2 \ \cdot \ 10 + 89 \ \cdot \ 3 + 5 \ \cdot \ 10 + 3$$
$$10^5 = 89(1 \ \cdot \ 10^3 + 1 \ \cdot \ 10^2 + 2 \ \cdot \ 10 + 3) + 5 \ \cdot \ 10 + 3 \tag{IV}$$

and

$$10^6 = 89(1 \ \cdot \ 10^4 + 1 \ \cdot \ 10^3 + 2 \ \cdot \ 10^2 + 3 \ \cdot \ 10 + 5) + 8 \ \cdot \ 10 + 5 \tag{V}$$

and this can be generalized to:

$$10^{n+1} = 89 \cdot (F_1 \cdot 10^{n-1} + F_2 \cdot 10^{n-2} + \dots + F_{n-1} \cdot 10 + F_n) + 10F_{n+1} + F_n \quad \text{(VI)}$$

for all positive integers n.

We then divide both sides by 10^{n+1} to get:

$$1 = \frac{89}{10^{n+1}} \cdot (F_1 \cdot 10^{n-1} + F_2 \cdot 10^{n-2} + \dots + F_{n-1} \cdot 10 + F_n) + \frac{10F_{n+1} + F_n}{10^{n+1}}.$$

We can establish that $\displaystyle\lim_{n \to \infty} \frac{10F_{n+1} + F_n}{10^{n+1}} = 0,$ the proof of which can be found in appendix B.

As we conclude our brief discussion of the eleventh Fibonacci number, let us take note of its successor: the twelfth Fibonacci number, $F_{12} = 144 = 12^2$. Aside from the first two numbers of the Fibonacci sequence, $F_1 = 1$, and $F_2 = 1$, it is the only Fibonacci number that is also a square number.

Displaying a Watch

What might the Fibonacci numbers have to do with the way a watch is displayed in a showcase or in an advertisement? Well, here we are using the geometric form of the Fibonacci numbers, namely, the golden rectangle. You may notice that advertisements and window displays often show a watch with the hands set at 10:10.

Figure 6-31

The hands make an angle of $19\frac{1}{6}$ minutes,[20] which is the equivalent of 155 degrees (figure 6-31). Now consider the rectangle formed by placing its vertices at points at which the hands indicate 10:10, that is, at hour markers 2 and $10\frac{1}{6}$. Then form a rectangle (figure 6-32) with these two points as two adjacent vertices and with the point of intersection of its diagonals at the center of the watch face. This rectangle is very close to the golden rectangle, whose diagonals meet at an angle of about 116.6 degrees (figure 6-33).

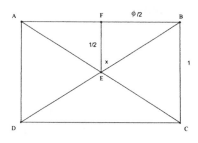

Figure 6-32   **Figure 6-33**

---

20. Since 10 minutes is $\frac{1}{6}$ of an hour, the hour hand moved $\frac{1}{6}$ of the distance between 10 and 11, from 10:00 to 10:10. Therefore, the hour hand moved $\frac{1}{6}$ of 5 minute markers, or $\frac{5}{6}$ of a minute marker. Thus the angle at 10:10 is $19\frac{1}{6}$ minute markers. To find the degree measure of that angle, we simply find the part of 60 that $19\frac{1}{6}$ is — $\dfrac{19\frac{1}{6}}{60} = \dfrac{\frac{115}{6}}{60} = \dfrac{115}{360}$, which is 115°.

This can be determined by using the golden rectangle (figure 6-33) and finding the measure of the angle whose tangent function is

$$\tan x = \frac{FB}{FE} = \frac{\frac{\phi}{2}}{\frac{1}{2}} = \phi$$

That angle, x, has measure 58.28 degrees. Therefore, $m\angle AEB = 116.56°$, which is very close to the 115° that is felt by many exhibitors to be the ideal placement of the hands of a watch. Some might have suspected that the 10:10 time placement was to allow the observer to see, in unobstructed fashion, the brand name of the watch. Not necessarily so. Consider an advertisement at the Prague airport (figure 6-34), where there was no need to highlight a brand name. Once again, the time 10:10 is shown.

Figure 6-34

Take, for another example, a recent United States postage stamp that exhibits the "American Clock."

Figure 6-35

Here the time shown is about 8:21½, where the hands form an angle of about **121.75°**,[21] which is not exactly the angle of the diagonals of an ideal golden rectangle. But it is close enough for the naked eye to appreciate. The golden rectangle seems to have again determined the ideal!

Some may recall that when they were taught to drive a car, they were instructed to hold the steering wheel at the positions of the 10 and 2 of a clock. Perhaps those points present some sort of special balance. Something to ponder!

---

21. To find the angle formed, we will do it in two parts. First, the angle formed between the "6," or 30-minute marker, and the 21.5-minute marker $\angle (30 - 21\frac{1}{2}) = 8.5 \cdot 6° = 51°$. Second, the angle formed by the 30-minute marker and the hour hand at 8:21.5. In 21.5 minutes the hour hand moved $\frac{21.5}{60}$ of the 5 minute markers between the 8 and 9 on the clock, or $\frac{21.5}{60}$ of 30°, which is 10.75°. The angle between the hour hand and the 30-minute marker is 10.75° + 60 ° = 70.75°. The angle formed by the hands of the clock is, then, 51° + 70.75° = 121.75°.

Fibonacci Helps Us Determine Seatings

There are often interesting mathematical phenomena that arise from problems that stem from seating arrangements. We will consider one here in which, once again, our Fibonacci numbers will emerge. Suppose that from a school full of boys and girls, you will be placing a number of chairs on a stage and you want to know how many ways to seat the boys and girls so that no two boys sit together. (This is not meant to be an indication that if two boys sit together, they will misbehave!) The chart in figure 6-36 will summarize the various seatings.

Number of Chairs	Arrangements	Number of Ways to Seat Students
1	B, G	2
2	BG, GB, GG, ~~BB~~	3
3	GGB, ~~BBG~~, BGB, GBG, GGG, ~~BBB~~, BGG, ~~GBB~~,	5
4	GGGG, ~~BBBB~~, ~~BBGG~~, ~~GGBB~~, BGBG, GBGB, ~~GBBG~~, BGGB, ~~BBBG~~, GGGB, ~~GBBB~~, BGGG, ~~BBGB~~, GGBG, ~~BGBB~~, GBGG	8
5	GGGGB, ~~BBBBB~~, ~~BBGGB~~, ~~GGBBB~~, BGBGB, ~~GBGBB~~, ~~GBBGB~~, ~~BGGBB~~, ~~BBBGB~~, ~~GGGBB~~, ~~GBBBB~~, BGGGB, ~~BBGBB~~, GGBGB, ~~BGBBB~~, GBGGB GGGGG, ~~BBBBG~~, ~~BBGGG~~, ~~GGBBG~~, BGBGG, GBGBG, ~~GBBGG~~, BGGBG, ~~BBBGG~~, GGGBG, ~~GBBBG~~, BGGGG, ~~BBGBG~~, GGBGG, ~~BGBBG~~, GBGGG	13

Figure 6-36

As you can clearly see, in each case the number of seating possibilities is a successive Fibonacci number, beginning with 2.

Let us now change the restrictions and insist that no boy or girl sits without someone from the same sex next to him/her *and* that the first position must be a girl. Figure 6-37 displays the various acceptable arrangements and shows us that the Fibonacci numbers give the number of ways this seating can be done. With six chairs, there are five possible seating arrangements.

Number of Chairs	Arrangements	Number of Ways to Seat Students
1	~~B, G~~	0
2	~~BG, GB~~, GG, ~~BB~~	1
3	~~GGB, BBG, BGB, GBG~~, GGG, ~~BBB, BGG, GBB,~~	1
4	GGGG, ~~BBBB, BBGG~~, GGBB, ~~BGBG, GBGB, GBBG, BGGB, BBBG, GGGB, GBBB, BGGG, BBGB, GGBG, BGBB, GBGG~~	2
5	~~GGGGB, BBBBB, BBGGB~~, GGBBB, ~~BGBGB, GBGBB, GBBGB, BGGBB, BBBGB~~, GGGBB, ~~GBBBB, BGGGB, BBGBB, GGBGB, BGBBB, GBGGB~~ GGGGG, ~~BBBBG, BBGGG, GGBBG, BGBGG, GBGBG, GBBGG, BGGBG, BBBGG, GGGBG, GBBBG, BGGGG, BBGBG, GGBGG, BGBBG, GBGGG~~	3

Figure 6-37

Fish in a Hatchery

Consider a fish hatchery that is partitioned into sixteen congruent regular hexagons—arranged in two rows (figure 6-38)—with pathways between each adjacent hexagon.[22] A fish begins a journey at the upper-left side of the double row of hexagons and ends up at the lower-right hexagon. Our problem is to determine the number of paths the fish can take to complete his journey to reach hexagon K, if he can only move to the right.

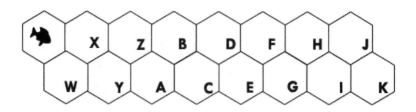

Figure 6-38

---

22. This example may not be too realistic, but it provides us with a nice application of the Fibonacci numbers.

As the fish begins his travels, there is only **one** path he can take to hexagon W (figure 6-39), since if he were to go horizontally first, he would no longer be traveling to the right to get to hexagon W.

Figure 6-39

To get to hexagon X, there are **two** paths, one directly to the right, and one through W (figure 6-40).

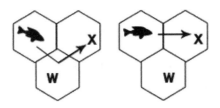

Figure 6-40

For the fish to get to hexagon Y, there are **three** paths: hexagons W-X-Y, X-Y, and W-Y (figure 6-41).

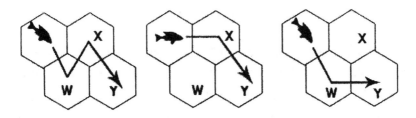

Figure 6-41

Perhaps you will anticipate the next number of possible hexagon paths that the fish can travel when there are five hexagons available. Yes, there are **five** paths, as shown in figure 6-42.

They are: W-Y-Z, W-X-Y-Z, X-Z, W-X-Z, and X-Y-Z.

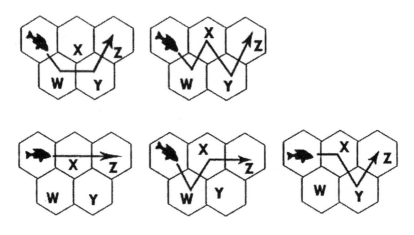

Figure 6-42

It would be correct to assume that this pattern will continue as we increase the number of hexagons for the fish to traverse. To summarize, we provide the following chart:

Number of hexagons for the fish's journey	1	2	3	4	5	6	7	...	n
Number of different paths	1	1	2	3	5	8	13		

Figure 6-43

So we might conclude that for the fish to reach the sixteenth hexagon, K, we would need to find F_{16}, which is 987. Thus, there are 987 possible paths for the fish to travel from the first hexagon to the sixteenth hexagon, K—always traveling only to the right.

The Game of Fibonacci Nim

At this point in the chapter, perhaps we shall have the enjoyment of playing a game where the winning strategy is dependent on the Fibonacci numbers. The rules of the game are as follows.

The Rules:

- There is one pile of pennies. The first player may remove any positive number of pennies, but not the whole pile.
- Thereafter, each player may remove at most twice the number of pennies his opponent took on the previous move.
- The player who removes the last penny wins.
- You move first: you can win from the initial position, but don't make any mistakes!

The questions before us, as we begin to play the game, are: Will the number of pennies at the beginning determine which player will win the game? What strategy should be used in playing the game? And what might this have to do with the Fibonacci numbers?

The strategy for this game, as you might have expected, is based on the Fibonacci numbers. Just as we can use binary notation to represent every positive integer as a sum of distinct powers of 2, we can use a similar process to write integers as a sum of Fibonacci numbers (see page 188). The winning strategy for this game involves removing the smallest of the Fibonacci summands.[23] Here is a method for doing this: pick the largest Fibonacci number not exceeding the given integer, then subtract that Fibonacci number and continue the process with the difference. For example, to write 20 in terms of Fibonacci numbers, subtract 13 from it to get 7. Subtract 5 from 7 to get 2, which itself is a Fibonacci number. Thus $20 = 1 \cdot 13 + 0 \cdot 8 + 1 \cdot 5 + 0 \cdot 3 + 1 \cdot 2 + 0 \cdot 1 = 101010$.[24]

---

23. A summand is one member of a sum.
24. We are merely representing the number by showing all the Fibonacci numbers: those for inclusion in the sum are multiplied by 1 and those not included in

We can express $20 = 13 + 5 + 2$, and thus the winning move involves taking two pennies. This is equivalent to taking the rightmost 1 in the Fibonacci representation of the number of pennies in the pile.

Number of pennies in pile	Fibonacci representation of the number of pennies	The smallest of the Fibonacci summands	Number of pennies that remain after the move	Fibonacci representation of the number of pennies that remain
20	101010	2	18	101000
19	101001	1	18	101000
18	101000	5	13	100000
17	100101	1	16	100100
16	100100	3	13	100000
15	100010	2	13	100000
14	100001	1	13	100000
13	100000	13	0	0
12	10101	1	11	10100
11	10100	3	8	10000
10	10010	2	8	10000
9	10001	1	8	10000
8	10000	8	0	0
7	1010	2	5	1000
6	1001	1	5	1000
5	1000	5	0	0
4	101	1	3	100
3	100	3	0	0
2	10	2	0	0
1	1	1	0	0

Figure 6-44

Compare the Fibonacci representations in the second column with those in the fifth column in figure 6-44. Note that the winning move in this game is equivalent to taking the rightmost 1 in the Fibonacci representation of the number of pennies in the pile. Less obvious is that the second player cannot make a similar move.

Try this strategy with a friend. Are there initial-pile sizes for which the first player cannot be assured of a win with this strategy?

---

the sum are multiplied by 0. Note: We do not use the first 1 of the sequence and begin with the second 1.

When Is a Fibonacci Number Equal to a Lucas Number?

What we seek to determine is when might a Fibonacci number, F_n, be equal to a Lucas number, L_m? Let's consider some of the relationships we have established earlier, which compare these two numbers.

For $m \geq 3$

$$L_m = F_{m+1} + F_{m-1} \geq F_{m+1} + F_2 > F_{m+1} \text{, and}$$

$$L_m = F_{m+1} + F_{m-1} < F_{m+1} + F_m = F_{m+2}.$$

In the first case, we have that $L_m > F_{m+1}$, and in the second case we get: $L_m < F_{m+2}$.

For $m \geq 3$, we have $F_{m+1} < L_m < F_{m+2}$.

Therefore, $F_n = L_m$ only when $m < 3$.

The only solutions are

$$F_1 = L_1 = 1$$
$$F_2 = L_1 = 1$$
$$F_4 = L_2 = 3$$

By this point in the book, you are probably expecting the Fibonacci numbers to appear, even where they may be least expected. The applications or sightings of these fabulous Fibonacci numbers seem to be boundless. Besides these kinds of appearances, the Fibonacci numbers can also surprise us by the many interrelationships they have, as well as those they have with the Lucas numbers. We will close this chapter with a sample listing of some of these amazing relationships. Convince yourself that they hold true by applying them to specific cases, or for the more ambitious reader, try to prove them true.

A Plethora of Fibonacci Relationships

Let k, m, n be any natural numbers; where k, m, $n \geq 1$:

$$F_1 = 1; \ F_2 = 1; \ F_{n+2} = F_n + F_{n+1}$$

$$L_1 = 1; \ L_2 = 3; \ L_{n+2} = L_n + L_{n+1}$$

$$11 \mid (F_n + F_{n+1} + F_{n+2} + \ldots + F_{n+8} + F_{n+9})$$

$(F_n, F_{n+1}) = 1$. The greatest common divisor of F_n and F_{n+1} is 1 (they are relatively prime)

$$\sum_{i=1}^{n} F_i = F_1 + F_2 + F_3 + F_4 + \ldots + F_n = F_{n+2} - 1$$

$$\sum_{i=1}^{n} F_{2i} = F_2 + F_4 + F_6 + \ldots + F_{2n-2} + F_{2n} = F_{2n+1} - 1$$

$$\sum_{i=1}^{n} F_{2i-1} = F_1 + F_3 + F_5 + \ldots + F_{2n-3} + F_{2n-1} = F_{2n}$$

$$\sum_{i=1}^{n} F_i^2 = F_n F_{n+1}$$

$$F_n^2 - F_{n-2}^2 = F_{2n-2}$$

$$F_n^2 + F_{n+1}^2 = F_{2n+1}$$

$$F_{n+1}^2 - F_n^2 = F_{n-1} \cdot F_{n+2}$$

$$F_{n-1} F_{n+1} = F_n^2 + (-1)^n$$

$$F_{n-k} F_{n+k} - F_n^2 = \pm F_k^2, \text{ where } n \geq 1 \text{ and } k \geq 1$$

$$F_m \mid F_{mn}$$

$$\sum_{i=2}^{n+1} F_i F_{i-1} = F_{n+1}^2, \text{ when } n \text{ is odd}$$

$$\sum_{i=2}^{n+1} F_i F_{i-1} = F_{n+1}^2 - 1, \text{ when } n \text{ is even}$$

$$\sum_{i=1}^{n} L_i = L_1 + L_2 + L_3 + L_4 + \ldots + L_n = L_{n+2} - 3$$

$$\sum_{i=1}^{n} L_i^2 = L_n L_{n+1} - 2$$

$$F_n F_{n+2} - F_{n+1}^2 = (-1)^{n-1}$$

$$F_{2k}^2 = F_{2k-1} F_{2k+1} - 1$$

$$F_{m+n} = F_{m-1} F_n + F_m F_{n+1}$$

$$F_n = F_m F_{n+1-m} + F_{m-1} F_{n-m}$$
$$F_{n-1} + F_{n+1} = L_n$$
$$F_{n+2} - F_{n-2} = L_n$$
$$F_n + L_n = 2F_{n+1}$$
$$F_{2n} = F_n L_n$$
$$F_{n+1} L_{n+1} - F_n L_n = F_{2n+1}$$
$$F_{n+m} + (-1)^m F_{n-m} = L_m F_n$$
$$F_{n+m} - (-1)^m F_{n-m} = F_m L_n$$
$$F_n L_m - L_n F_m = (-1)^m 2 F_{n-m}$$
$$5F_n^2 - L_n^2 = 4(-1)^{n+1}$$
$$F_{n+1} L_n = F_{2n+1} - 1$$
$$F_n - F_{n-5} = 10 F_{n-5} + F_{n-10}$$

$$3F_n + L_n = 2F_{n+2}$$

$$5F_n + 3L_n = 2L_{n+2}$$

$$F_{n+1}L_n = F_{2n+1} + 1$$

$$L_{n+m} + (-1)^m L_{n-m} = L_m L_n$$

$$L_{2n} + 2(-1)^n = L_n^2$$

$$L_{4n} - 2 = 5F_{2n}^2$$

$$L_{4n} + 2 = L_{2n}^2$$

$$L_{n-1} + L_{n+1} = 5F_n$$

$$L_m F_n + L_n F_m = 2F_{n+m}$$

$$L_{n+m} - (-1)^m L_{n-m} = 5F_m F_n$$

$$L_n^2 - 2L_{2n} = -5F_n^2$$

$$L_{2n} - 2(-1)^n = 5F_n^2$$

$$L_n = F_{n+2} + 2F_{n-1}$$

$$L_n = F_{n+2} - F_{n-2}$$

$$L_n = L_1 F_n + L_0 F_{n-1}$$

$$L_{n-1}L_{n+1} + F_{n-1}F_{n+1} = 6F_n^2$$

$$F_{n+1}^3 + F_n^3 - F_{n-1}^3 = 3F_{3n}$$

Chapter 7

The Fibonacci Numbers Found in Art and Architecture

The golden ratio seems to be ever present in both art and architecture. Why this is, we can only speculate. We have already established that it has been proclaimed the most beautiful ratio of all time—both numerically and geometrically. Here, we will provide a brief survey of its apparent sightings in these two aesthetic fields. And we will look at the visual art in both two and three dimensions.

During the Renaissance, the study of proportions in architecture, sculpture, and painting as an aesthetic mathematical device was based on the notion that harmony and beauty in art can best be determined through particular numbers, often seen in their relationships with other numbers. This is how the golden section came into prominence. There are countless articles and books that attempt to show that outstanding art and architecture derived from, in one form or another, the principle of the golden section.

The golden section, whose ratio is also considered beautiful, was already known in antiquity and by the Pythagoreans when they constructed a regular pentagon and a regular dodecahedron. Many architects through the ages, intuitively or deliberately, used the golden section in their sketches and construction plans, either for the entire work or for the apportionment of parts of the work. In particular, this aesthetic bias frequently manifested itself as the golden rectangle. Since the golden ratio, $\phi = \frac{\sqrt{5}+1}{2}$, is an irrational number, architects were known to use lengths of the sides of a rectangle whose ratio approximated ϕ. And, as you now know, the most common approximations for ϕ are the Fibonacci numbers. Moreover, the larger these numbers, the closer the ratio is to ϕ.

Fibonacci Numbers in Architecture

For years, architects have used the golden ratio, ϕ, in their drawings, sometimes as the golden rectangle, other times as a partition indicator. To circumvent the difficulty of using the irrational number (ϕ), the golden rectangles that were constructed typically involved dimensions that were Fibonacci numbers. The golden ratio—an ideal—was very nearly achieved, since, as we already know, the ratio of two consecutive Fibonacci numbers approaches the golden ratio, with the greater the magnitude of the Fibonacci numbers, the closer the approximation of ϕ.

Perhaps the most famous structure that exhibits the golden rectangle is the Parthenon on the Acropolis in Athens. The artistic concept for building this magnificent structure came from Phidias (460–430 BCE). It was erected to thank Pericles (500–429 BCE) for saving Athens during the Persian War (447–432 BCE). The purpose of this temple was to house the statue of the goddess Athena. It is believed that the current designation for the golden ratio, ϕ, came from the first letter of Phidias's name (in Greek: ΦΣΙΔΙΑΣ). We must stress at this point that we have no evidence today that Phidias was aware of the golden ratio; we can only assume it from the shape of the structure. As you can see in figure 7-1, the Parthenon fits nicely into a golden rectangle. Furthermore, in figure 7-1, you will notice a number of golden rectangles, generated from the structure of the building.

Figure 7-1

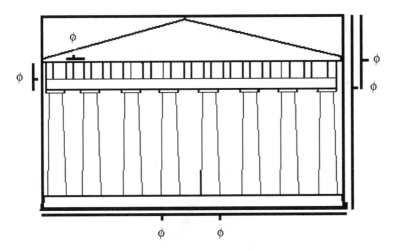

Figure 7-2

If one takes measurements of various ornaments above the columns, the golden ratio keeps cropping up. Figure 7-2 exhibits the golden ratio throughout the Parthenon. To what extent ancient architects consciously used the golden rectangle is still largely a mystery. Some believe that we moderns are just looking for ways

to superimpose a golden rectangle over a specific structure—
though no one can deny its presence.

Another example of the golden section can be seen on
Hadrian's Arch on the Propylaeum temple, also situated on the
Acropolis. The arrangement of snails and other ornaments at the
top of the arch further exhibits the golden ratio through a regular
pentagon that can be superimposed as in figure 7-3. (Notice the **M**
represents the major portion of the golden section, while the **m** represents the minor portion.)

Figure 7-3

Figure 7-4
(Copyright Wolfgang Randt. Used with permission.)

According to the Greek historian Herodot (ca. 485–424 BCE), the Khufu (Cheops) Pyramid at Giza was constructed in such a way that the square of the height is equal to the area of one of the lateral sides. Using our superimposed triangle in figure 7-4 (and the Pythagorean theorem), we get the following:

$$h_\Delta^2 = \frac{a^2}{4} + h_P^2$$

The area of one of the lateral triangles is $A = \frac{a}{2} \cdot h_\Delta$.

According to the construction principle introduced by Herodot (above), where the square of the height is equal to the area of the lateral triangle, we get:

$$h_P^2 = h_\Delta^2 - \frac{a^2}{4} = A = \frac{a}{2} \cdot h_\Delta.$$

If we divide both sides of the equation $\frac{a}{2} h_\Delta = h_\Delta^2 - \frac{a^2}{4}$ by $\frac{a}{2} h_\Delta$,

we get:

$$1 = \frac{h_\Delta}{\frac{a}{2}} - \frac{\frac{a}{2}}{h_\Delta}$$

By making the following substitution of $\dfrac{h_\Delta}{\frac{a}{2}} = x$ by first changing

its form to $\dfrac{\frac{a}{2}}{h_\Delta} = \dfrac{1}{x}$, we get the equation: $1 = x - \dfrac{1}{x}$, which just happens to generate the equation $x^2 - x - 1 = 0$, whose solution is already known to us as $x_1 = \phi$ and $x_2 = -\dfrac{1}{\phi}$. By now you have realized

that x_2 is negative and holds no real meaning for us geometrically; so we won't consider it here. Using today's measurement capabilities, this great pyramid has the following dimensions.

Cheops pyramid	Length of the side of the base: a	Height of lateral triangle: h_Δ	Pyramid height h_P	$\dfrac{h_\Delta}{\dfrac{a}{2}}$	$\dfrac{C}{2h_P}$
measurements	230.56 m	186.54 m	146.65 m	1.61813471	3.144357313

Figure 7-5

Lo and behold, the ratio of the height of the lateral triangle to half its base is $\dfrac{h_\Delta}{\dfrac{a}{2}} = 1.61813471$.

Was this done intentionally by the design of an ingenious architect? This is a question that you'll have to answer for yourself. We can only point out what has been found by measurement and historical clues.

We assume that the Egyptians did their measurements with cubits, which was defined as the length of a man's forearm (i.e., from fingertip to elbow) and measured as a length of 52.25 cm. A British Egyptologist, W. M. F. Petrie (1853–1942), established measurements for this famous pyramid as follows:

$$\frac{a}{2} = 220 \text{ cubits } (\approx 114.95\text{m}) \text{ [i.e. } a = 440 \text{ cubits} \approx 229.90\text{m]}$$

$$h_p = 280 \text{ cubits } (\approx 146.30\text{m})$$

$$h_\Delta = 356 \text{ cubits } (\approx 186.01\text{m})$$

The angle formed by the lateral height (h_Δ) and the base yields a cosine measure, which is a quite-surprising result: $\cos\angle(\frac{a}{2}, h_\Delta)$ $= \dfrac{220}{356} = \dfrac{55}{89}$. You would almost think that there is some sort of "fix" in this analysis. However, we are just reporting the facts as they have been recorded.

From the above, we can speculate that the base of the pyramid, whose side is 230.56 m, has a perimeter of 922.24 m. If we divide this perimeter by twice the height of the pyramid, we get, curiously enough, an approximation for one of the most famous numbers in mathematics, π.[1]

Spectators have noted the golden section on Mexican pyramids, Japanese pagodas, and even Stonehenge (ca. 2800 BCE) in England. There is, moreover, speculation that the Arch of Triumph in ancient Rome was built on the basis of the golden ratio.

The oldest stone structure in Germany, stemming back to the post-Roman period (770 CE), is the "Königshalle" in the town of Lorsch (figure 7-6). This magnificent example of architecture from the early Middle Ages has an impressive open-air ground floor. The interior space of this building exhibits practically a perfect golden rectangle. We still don't know today what purpose this building had and how it was used. Perhaps if one day the building plans could be found, we might be able to see if the dimensions might have intentionally been Fibonacci numbers.

Figure 7-6

The Cathedral of Chartres, France (built from 1194–1260), clearly exhibits the golden rectangle. In figure 7-8, you can see the front portal, and figure 7-7 shows a window from the Cathedral of Chartres exhibiting this famous rectangle.

---

1. For more on π, see Alfred S. Posamentier and Ingmar Lehmann, π: *A Biography of the World's Most Mysterious Number* (Amherst, NY: Prometheus Books, 2004).

Figure 7-7   **Figure 7-8**

Figure 7-9

Figure 7-10

Throughout the Renaissance, the designs of many construction projects used the Fibonacci numbers or the golden ratio. They can be found in the plans of the dome of the Santa Maria del Fiore Cathedral in Florence (see figure 7-9). The rough sketch by Giovanni di Gherardo da Prato (1426) exhibits the Fibonacci numbers, 55, 89, and 144, as well as 17 (which is half the Fibonacci number 34) and 72 (which is half the Fibonacci number 144). The dome was actually constructed in 1434 by Filippo Brunelleschi (1337–1446) and has a height of 91 m and a diameter of 45.52 m, which (sadly) does not give us the golden ratio.

The remainder of the Cathedral in Florence further demonstrates the Fibonacci numbers. But the most dramatic proportions are those shown in figure 7-11, where the ratio of 89:55 (= 1.6181818...) has an extremely close approximation to the golden ratio 1.618033988... .

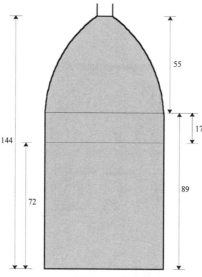

Figure 7-11

Let's move to more modern times. Le Corbusier (1887–1965),[2] was a French-Swiss architect who between 1946 and 1952 designed the *unités d'habitation* in Marseille, France (see figure 7-12), which again exhibits the golden ratio. Here it is exhibited in the tower, which divides the remainder of the building into the golden ratio. This was not done by chance, but rather by specific design! He displays the golden section throughout.

Figure 7-12

2. His actual name was Charles-Édouard Jeanneret.

Furthermore, all of the measurements in the apartments are based on a theory of the proportions that Le Corbusier developed in 1948, and published in a book titled *The Modulor: A Harmonious Measure to the Human Scale Universally Applicable to Architecture and Mechanics.*[3] This stringent rule caused somewhat of a furor in the art world. Le Corbusier developed his scheme on the basis of the golden section. This included his decision that the height of a door was to have a measure of 2.26 m (7.4 feet),[4] so that a person of height 1.83 m (6 feet) can touch the top with outstretched arms. Le Corbusier supported his ideas with the following: "A man with a raised arm provides the main points of space displacement—the foot, solar plexus, head, and fingertip of the raised arm—three intervals which yield a number of golden sections that are determined by Fibonacci."

The *Modulor* (see figure 7-13), a model of how the human proportions should be standardized for use by architects and engineers, consists of two scales or bands, which are marked with two sequences of natural numbers—the sequence:

(*)   6, 9, 15, 24, 39, 63, 102, 165, 267, 432, 698, 1130, 1829

and an intermediate sequence:

(**)   11, 18, 30, 48, 78, 126, 204, 330, 534, 863, 1397, 2260

Usually, at second glance, we can see the relationship to the golden section, ϕ.

3. Le Corbusier, *The Modulor: A Harmonious Measure to the Human Scale Universally Applicable to Architecture and Mechanics and Modulor*, 2 vols. (Basel: Birkhäuser, 2000).
4. 1 meter = 3.28 feet.

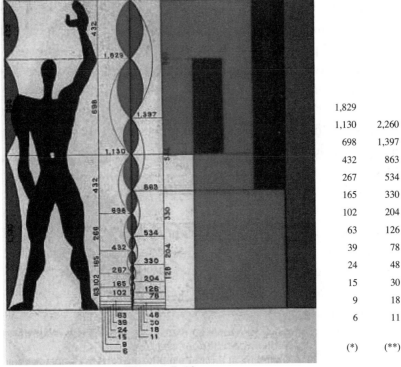

1,829	
1,130	2,260
698	1,397
432	863
267	534
165	330
102	204
63	126
39	78
24	48
15	30
9	18
6	11
(*)	(**)

Figure 7-13

If we divide the members of the sequence (*) by 3 and those of the sequence (**) by 6, we get:

(*)′ 2, 3, 5, 8, 13, 21, 34, 55, 89, 144, $232\frac{2}{3}$, $376\frac{2}{3}$, $609\frac{2}{3}$

and

(**)′ $1.8\overline{3}$, 3, 5, 8, 13, 21, 34, 55, 89, $143.8\overline{3}$, $232.8\overline{3}$, $376.8\overline{3}$

If we round off the fractional members of each sequence, we are then dealing exclusively with the Fibonacci numbers.

This relationship to the Fibonacci numbers results from the fact that Le Corbusier's sequences almost approximate a geometric progression with the common ratio of the golden section, ϕ. In his *Modulor,* Le Corbusier claimed to have found a scale that adapted exceptionally well to the visual arts.

Le Corbusier was substantially influenced by the ideas of the Romanian lawyer, engineer, and diplomat Matila Costiescu Ghyka

(1881–1965),[5] whose writing about the golden ratio connected the religious aspects of Luca Pacioli (1445–1517)[6] with the aesthetic writings of Adolf Zeising (1810–1876).[7] Ghyka interpreted the golden section as a fundamental secret of the universe and supported this position with examples from nature. This perspective further inspired Le Corbusier, whose aim was to create a harmonic design over a specific living space. He seized on the same philosophy in creating resident apartments.

Figure 7-14

In 1947 Le Corbusier was a member of the architectural commission that planned the building of the UN headquarters in New York (see figure 7-14). He was ultimately responsible for the basic concept of the plans for the thirty-nine-story Secretariat building. As you might expect, the height and the width of the building are close to the ratio of 1.618:1. Yes, once again the golden ratio stares us in the face!

Last, but not least, an obvious exhibit of the golden ratio can be found in an aerial view of the Pentagon in Washington, DC (see

5. Matila Ghyka, *Esthétique des proportions dans la nature et dans les arts* (Paris, 1927); Matila Ghyka, *The Geometry of Art and Life* (New York: Sheed and Ward, 1946; reprint, New York: Dover Science Books, 1977).
6. Fra Luca Pacioli, *Divina Proporzione—A Study of the Golden Section First Published in Venice in 1509* (New York: Abaris Books, 2005).
7. Adolf Zeising, *Neue Lehre von den Proportionen des menschlichen Körpers* (Leipzig: R. Weigel, 1854); the book *Der goldene Schnitt* was printed posthumously at the expense of the Leopoldinisch-Carolinischen Akademie: Halle, 1884.

figure 7-15). By now you might (rightly) expect that the intersection point of two diagonals divides each of the diagonals in the golden ratio, as is the case with any regular pentagon.

Figure 7-15

In the field of architecture, there are countless illustrations of the use of the golden ratio, which, of course, employ the Fibonacci numbers. The lingering question persists: Are these coincidences or are these golden sections marked off deliberately? In the case of some buildings such as those designed by Le Corbusier, we have enough evidence to know that the use of the golden ratio was intentional. As for the ancient architecture, we can sometimes speculate or assume that if an architect had a connection with a mathematician, then he might have used it by design. Still, there is always the notion that this ratio delivers the most beautifying partition in geometry and therefore the "trained eye" might have simply landed on the golden ratio through artistic prowess.

Fibonacci Numbers and Sculptures

The golden section, ϕ, endows a work of art with a particularly pleasing and "beautiful" shape. The golden section, ϕ, also serves as the bridge to the Fibonacci numbers. Recall that the ratio of consecutive members of the Fibonacci sequence approximates the golden section:

$$\frac{5}{3} = 1.666\ldots, \quad \frac{8}{5} = 1.6, \quad \frac{13}{8} = 1.625, \quad \frac{21}{13} = 1.615\ldots$$

The roughest approximation occurs with the smallest Fibonacci numbers, and yet even then its greatest deviation from the actual value ϕ is less than $\frac{1}{20}$, or 5 percent.

In 1509 the Italian Renaissance scholar Fra Luca Pacioli (ca. 1445–1517), in his book on regular solids, titled *De divina proportione*,[8] stressed the importance of the golden section and enthusiastically called it "De Divina Proportio" (i.e., divine proportion). The use of the adjective *divina* shows the level of aesthetic significance he attached to the golden section.

In the first part of the twentieth century, two important books on the subject were published: in 1914 *The Curves of Life*[9] written by Sir Theodore Andrea Cook, and in 1926 *The Elements of Dynamic Symmetry* by Jay Hambidge.[10] Both books had a great effect on artists, in particular the latter work. Hambidge's primary discovery was that there were two types of symmetry—static and dynamic. His thorough study of Greek art and architecture, particularly of the Parthenon, convinced him that the secret of the beauty of Greek design was in the conscious use of dynamic symmetry—the law of natural design based upon the symmetry of growth in man and in plants.

The scholar Otto Hagenmaier[11] pointed out that the Apollo Belvedere (a Roman copy of Leochares; Vatican Museum, Rome)[12] is an example of a famous sculpture with proportions that reflect the golden section. In figure 7-16 you can see how the sculpture measures up in its golden ratio proportions.

---

8. Luca Pacioli, *De divina proportione* (Venezia, 1509; reprinted in Milan in 1896, and Gardner Pelican, 1961).

9. (London, 1914; reprint, New York: Dover, 1979).

10. (New York, Dover, 1967). The dynamic symmetry is based on the laws of thermodynamics.

11. Otto Hagenmaier, *Der Goldene Schnitt—Ein Harmoniegesetz und seine Anwendung* (Gräfeling, Germany: Moos & Partner, 1958); 2nd ed. (München: Moos Verlag, 1977).

12. Attic sculptor of the fourth century BCE; the Roman copy was found at excavations in Rome toward the end of the fifteenth century.

Figure 7-16
Roman copy of *Apollo*
Belvedere.

Figure 7-17
Aphrodite of Melos.

The great mathematician Felix Klein (1849–1925) could not rest until he tried to explain the beauty of this magnificent sculpture mathematically and beyond the golden ratio. He marked the significant points of a Gaussian curve[13] on the face of Apollo in order to discover the secret of its beauty. Is this really what we expect from mathematics? Maybe not, but it opens our eyes wider to the mysteries of aesthetics.

The *Aphrodite of Melos* (also called *Venus de Milo*) in the Louvre museum in Paris was created at about 125 BCE (see figure 7-17). Her navel partitions the entire figure into the golden ratio. Was this by chance or by design? We will never know for certain.

---

13. The Gaussian curve is the normal curve.

The artists' association *Section d'Or* was founded in 1912 by Jacques Villon (1875–1963)[14] in Paris, during the late Cubism period. The association's name itself already highlights its belief in the golden section. Its stated goal was to create absolute painting purely from numerical ratios in which the golden section was not just nominally used, but rather was the theoretical starting point. The French-Hungarian sculptor and painter Étienne Béothy (1897–1961) did not confine himself to the two-dimensional painting that the *Section d'Or* community did, but expanded his work to the three-dimensional realm. Béothy supported his theoretical groundwork further by providing a detailed construction plan for all his sculptures. He relied in large measure on the "golden series"—apparently a reference to the Fibonacci numbers—for the distances, curves, and other proportions in his works.

In 1926 Béothy wrote in the preface to the Hungarian version of his book *La Serie d'Or,* [15] "It [the *golden series*] has the same significance for the visual arts as harmony has for music. . . . What works for music will also be achieved for the visual arts through the golden series." At the end of this book, Béothy concludes that the best sequence of numbers that can represent the golden series is the additive sequence:

$$\ldots 0, 1, 1, 2, 3, 5, 8, 13, 21, 34, 55, 89, 144, \ldots^{[16]}$$

14. Whose actual name was Gaston Duchamp, French painter and graphic artist.
15. Uwe Rüth, Etienne Beothy, and Helga Müller-Hofstede, *Etienne Beothy: Ein Klassiker der Bildhauerei-Retrospektive* (Sculpture Museum, Marl, Germany, Exhibition 1979, ed. Dr. Uwe Rüth).
16. Étienne Béothy, *La Serie d'Or* (Paris : Edition Chanth, 1939).

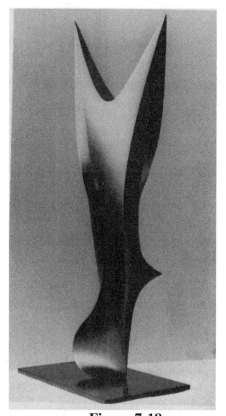

Figure 7-18
(Used with permission of Uwe Rüth.)

Béothy's wood sculpture, *Essor II* (op. 77), composed in 1936 (Sculpture Museum, Marl, Germany), appears to grow from a narrow footing and twists upward (see figure 7-18). It appears as if it were a flame. Yet behind this sculpture lie exact plans based on the Fibonacci numbers. It is out of this mathematical basis that the sculptor derived a work of great beauty.

Through the minimalist art movement in the 1960s the use of the Fibonacci numbers became fashionable. Some works by Timm Ulrichs (b. 1940) rely entirely on the golden section. For a sculpture (figure 7-19), he cut bread and vegetables in the proportion of this number, ϕ. He also sliced up a book with the title *The Golden Section* by Otto Hagenmaier[17] into the golden section!

---

17. *Der Goldene Schnitt: Ein Harmoniegesetz und seine Anwendung,* 2nd ed. (Gräfeling, Germany: Moos & Partner, 1958).

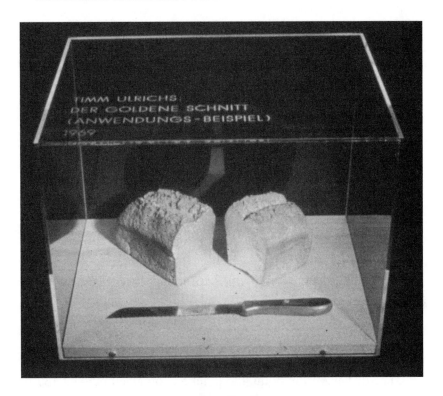

Figure 7-19
Timm Ulrichs, *Der goldene Schnitt* (Application example),
1969.
(Photo by Timm Ulrichs. Used with permission.)

In 1969 Ulrich wrote "I took several food products—bread, sausage, and pickles that one usually divides with a knife—and deliberately partitioned them into the golden section, and then emblazoned it in gold."

Jo Niemeyer (b. 1946), a German graphic artist and architect, created works according to plans that were based on the golden section. He even called his works *Geometric Compositions*. Niemeyer is still moving on—in the true sense of the word! His work, *20 Steps around the Globe*[18] was a breathtaking "country-art" project that he completed in 1997. Once again, the golden section is the key element of the project. It involved the installation of twenty high-grade

---

18. Ulrich Grevsmühl, *20 Punkte—20 Steps. Ein Land-Art Projekt von Jo Niemeyer*. See *20 Steps around the Globe*. http://www.jo.niemeyer.com.

steel columns once around the earth. The columns were placed at carefully calculated points along a great circle. [19] These twenty points were determined by dividing the great circle around the earth, whose circumference was given as 40,023 km. Yet the golden ratio played a crucial role in determining the position of the points.

Figure 7-20

The distance from point 1 to point 2 was determined to be .458 m (see figure 7-20).

Then the distance of point 2 to point 3 was determined by $\phi \cdot .458$ m = .741 m. Point 3 is then 1.198 m from point 1.

The distance from point 3 to point 4 was determined by $\phi \cdot 1.198$ m = 1.939 m. That means that point 4 is 3.137 m from point 1.

The distance from point 4 to point 5 is obtained by $\phi \cdot 3.137$ m = 5.077 m. Then point 5 is 8.214m from point 1.

By continuing in this way, he found that point 20 would coincide with point 1 (with all points along a great circle).

The specific positions of the columns were more exactly carried out with the help of the golden angle [20] (see figure 7-21).

The first column was placed in Lapland (Finland). This is near the Swedish border and north of the polar circle. The eighth column can be seen on the photo of the starting point (figure 7-22).

The complete route (figure 7-21) is described with all necessary data in the project *20 Steps around the Globe*[21] at the Web site http://www.jo.niemeyer.com. Jo Niemeyer believes that only the proportions are essential for the perception and understanding of this feat, and says, "There are no measures—only the proportions exist."

---

19. The great circle of a sphere is the largest circle that can be drawn on a sphere and has its center at the center of the sphere.

20. The golden angle (see p. 148) ≈ 137.507 764 050 . . .°.

21. The geodesic draft was carried out under the scientific advice of Dr. Ulrich Grevsmühl (Hinterzarten); the computer program was by Jörg Pfeiffer. See *20 Steps around the Globe*, http://www.jo.niemeyer.com.

Positions

1-12 Ropinsalmi Lapland Finland
    0-18 km
13 Kautokeino District Norway
    47 km
14 Anarjokka Norway
    124 km
15 Pitkajarvi/Nickel Russia
    325 km
16 The Barents Sea Int. Water
    851 km
17 Noryj Urengoj Siberia Russia
    2,229 km
18 Baotou Inner/Mongolia China
    5,837 km
19 Westernport Bay Australia
    15,282 km
20 = 1 Ropinsalmi Lapland Finland

Figure 7-21

Figure 7-22
The starting point: Ropinsalmi/Lapland/Finland,
Position 68°40'06"N 21°36'21"E.
(Reprinted with permission of Jo Niemeyer.)

The Italian artist Mario Merz (1925–2003) set up a monument to the Fibonacci numbers in several works of art. He is the doyen of the arte povera (the poor art), which distinguishes itself by the meager materials and simple gestures it employs. It is the European answer to the American minimalist art. For him the Fibonacci numbers were a guiding light as well as an inspiration.

His works *Fibonacci Napoli* (1971) and *Animali da 1 a 55* (1997–2000) are his best known. Yet, as an example of his work in a black-and-white format, his drawing *Ivy* may more graphically illustrate his respect for the simple elegance of the Fibonacci numbers (see figure 7-23).

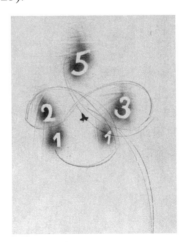

Figure 7-23

By reaching out to the Fibonacci numbers, Merz relegates to nature a reason for its power. "The Fibonacci numbers are in a rapid expansion. . . .The numbers have the power, to ease walls." This is how Mario Merz describes the peculiarity and effect of this historical design system.[22]

Figure 7-24
(Reprinted with permission of Kunstmuseum Wolfsburg.)

On eight marble plates, Merz assembled his sculpture *Igloo Fibonacci* (1970) with the help of brass pipes and steel hinges; on this he mounted adhesive pieces of tape with (white) numbers. The eight arms of the igloo are connected to each other with joints at the distances that reflect the Fibonacci proportions (see figure 7-24.)

Obviously Merz was quite fascinated with the Fibonacci numbers, as is further evidenced by his having decorated the great hall in the Zurich, Switzerland, central railway station with the Fibonacci numbers. He also used the Fibonacci numbers to decorate smokestacks: the 52 m smokestack at the Center for International Light Art in Unna, Germany, in 2001 (figure 7-25), and a smokestack in Turku, Finland, in 1994 (figure 7-26).

---

22. Mario Merz, *Die Fibonacci-Zahlen und die Kunst*, in Mario Merz's Exhibition catalogue of the Folkwang-Museum, Essen, Germany, 1979, p. 75.

Figure 7-25   **Figure 7-26**

The Austrian artist Hellmut Bruch (b. 1936) was also greatly influenced by the Fibonacci numbers as evidenced by his many designs that clearly reflect the Fibonacci numbers in the relative lengths of their components (see figures 7-27, 7-28, and 7-29.) It is said that these lengths enabled him to create the most astounding shapes of tangible and evident harmony.

The columns, made of stainless steel, exhibit through their lengths a portion of the Fibonacci sequence (89 cm, 144 cm, 233 cm, 377 cm, 610 cm). The name *Hommage à Fibonacci* is an eloquent testimonial.

Employing the Fibonacci sequence in nearly all of his work, Bruch has said that through these numbers he tries to show the fullness of forms and the variety of nature and tries to understand them. Bruch is not only interested in demonstrating the Fibonacci series but, more important, in seeing what these numbers can do to enhance his artistic possibilities.

Figure 7-27   **Figure 7-28**   **Figure 7-29**
(Reprinted with permission of Hellmut Bruch.)

The German sculptor Claus Bury (b. 1946) heralds the theme of his sculpture merely with its name: *Fibonacci's Temple*. The structure, built in 1984 and located in Cologne, Germany, is made of raw spruce wood and is 15 m long, is 6.3 m wide, and 5.70 m high at the center (figure 7-30).

Figure 7-30
(Reprinted with permission of Claus Bury.)

The preparation for this structure reveals that Bury was already fascinated with the Fibonacci numbers and in particular with the theoretical writings of Le Corbusier in *Modulor* (see pages 241–42). An inspection of Bury's plans demonstrates that he clearly constructed this work according to the Fibonacci numbers (see figures 7-31 and 7-32).

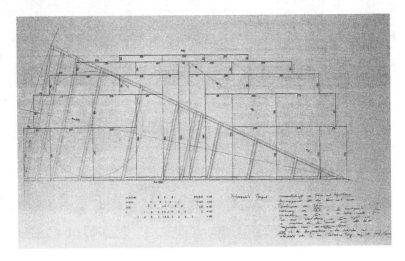

Figure 7-31
(Reprinted with permission of Claus Bury.)

Figure 7-32
(Reprinted with permission of Claus Bury.)

The Fibonacci numbers clearly have inspired modern artists in their sculptures. Yet we still remain uncertain about the conscious influence that these fabulous numbers have had on the ancient artists. Were they aware of these numbers or of the golden ratio, or did they do these works intuitively and they just happened to fit the golden ratio? This leads us to look for further answers in the two-dimensional medium of painting.

Fibonacci Numbers and Paintings

Leonardo da Vinci (1452–1519)[23] illustrated the book *De divina proportione*[24] by the Franciscan monk Fra Luca Pacioli (ca. 1445–1517) with an anatomical study of the "Vitruvian" man. This drawing supported Pacioli's discussion of the Roman architect Vitruvius (ca. 84 BCE–27 BCE)[25] whose documents about architecture suggest that he believed the proportions of the human body are a basis for architecture. The ratio of the square side to the circle radius in this well-known picture (figure 7-33) corresponds to the golden section with a deviation of 1.7 percent.

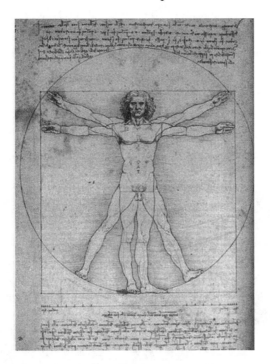

Figure 7-33

23. Famous Italian painter, sculptor, architect, natural scientist, and engineer.

24. Luca Pacioli, *De divina proportione* (Venezia, 1509; reprinted in Milan in 1896, Gardner Pelican, 1961).

25. Actually Vitruvius Pollio, Roman armed forces technician and engineer. He published ten books about architecture and civil engineering and developed a theory of proportions.

The respected German painter and graphic artist Albrecht Dürer (1471–1528) also relied heavily on Vitruvius's work when he wrote his *Four Books on Human Proportions* in 1523. Here he refines the earlier works and expresses them as a system of proportions, where the unit of measure is the human body with the parts expressed as fractions thereof.

In Dürer's books on geometry, fortress building, and human proportions, he supported all of his ideas with theoretical (mathematical) arguments, as opposed to aesthetic ones. As a matter of fact, his works in descriptive geometry influenced some of the greatest thinkers of the Renaissance—Johannes Kepler (1571–1630) and Galileo Galilei (1564–1642) among them.

In his famous self-portrait (ca. 1500), Dürer drew himself with a head of wavy hair, the outlines of which create an equilateral triangle. This can be seen in figure 7-34, where Dürer actually superimposes the triangle and several other guidelines over the self-portrait. The base of the equilateral triangle divides a height of the entire picture into the golden ratio. The chin also divides the height of the entire picture into the golden ratio (in the other direction).

Figure 7-34

It remains unclear why Dürer never connected the golden section to his drawing of the regular pentagon,[26] since at age twenty-nine he had constructed the golden section in his self-portrait.

Another classic work of art is Sandro Botticelli's painting *The Birth of Venus* (1477). Sir Theodore A. Cook (1867–1928) analyzed it with the aid of the golden section. He superimposed a number scale on the figure of Venus, which had the first seven powers of ϕ (see figure 7-35).

As earlier, in our review of architectual and scultpural applications of the Fibonacci numbers, we are faced here with a similar problem to determine whether conscious use was made of the golden section or the Fibonacci numbers. Many scholars collected information in the nineteenth and twentieth centuries to search for applications of the golden section. These applications need to be critically and sensitively inspected. Oftentimes, we stumble on fractions that approach the golden section and immediately assume that the golden section was intended. We must be careful not to draw conclusions where they aren't necessarily warranted.

Figure 7-35

26. See page 155ff for a discussion of Dürer's construction of the "regular pentagon," one that looks regular but is just slightly inaccurate.

Let us analyze a few samples where the golden section appears, though you could unearth countless examples from the past several centuries in which you could find similar "sightings." We will begin with, perhaps, the most famous painting in Western civilization, Leonardo da Vinci's *Mona Lisa* (figure 7-36), which was painted between 1503 and 1506 and is on exhibit in the Louvre museum in Paris. King François I of France paid 15.3 kilograms of gold for this painting. Let's look at this masterpiece from the point of view of the golden ratio. First, one can draw a rectangle around Mona Lisa's face and find that this rectangle is a golden rectangle. In figure 7-37, you will notice several triangles; the two largest ones are golden triangles. Furthermore, in Figure 7-38, you will observe specific points on the body of Mona Lisa as golden sections. Since da Vinci illustrated Pacioli's book *Da divina proportione*, which thoroughly discussed the golden section, it can be assumed that he was consciously guided by this magnificent ratio.

Figure 7-36   **Figure 7-37**   **Figure 7-38**

Another great painter Raphael (1483–1520) also exhibits the golden section in his *Sistine Madonna* (1513). The horizontal line indicated in figure 7-39, which emanates from the eyes of Pope Sixtus II and the head of Saint Barbara, divides the height of the picture into the golden section and also partitions the figure of the Madonna into two equal parts. In figure 7-40, you will notice that the lines emanating from very selected points in the figure aid in determining the golden ratio from another point of view; here, too, the Madonna's figure is divided by the golden section. You will also observe that the isosceles triangle, the ratio of whose base and altitudes closely approximates the golden ratio, superimposed on the picture in figure 7-40 exactly encases the four heads. Whether Raphael consciously selected these dimensions with the golden section in mind, or whether these were merely the product of an artistic eye, remains the master's secret.

Figure 7-39

Figure 7-40

In figure 7-41 a regular pentagram has been superimposed over Raphael's *Madonna Alba* (1511–1513). This geometric form is clearly visible along appropriate linear parts. This once again demonstrates Raphael's penchant for the golden section—which we already know is embedded in the regular pentagram.

Figure 7-41
(National Gallery, Washington, DC.)

Let's jump a few hundred years ahead to the French painter Georges Seurat (1859–1891), who was known for having a strict geometric structure to his works. Although he used colors to a dramatic extent, Seurat wanted to highlight his use of geometric space as the main attraction, which thereby helped set the groundwork for modern art.

Figure 7-42

The painting *Circus Parade* (1888, New York, Metropolitan Museum of Art) by Seurat (figure 7-42) is replete with evidence of the golden section. Some observers see the golden section produced by the horizontal line at the top edge of the banister, and in the vertical direction by the line that is just to the right of the main figure. Others find that a golden rectangle is formed by these two lines and the horizontal line just below the nine lights at the top of the picture. There is also speculation that the golden spiral is embedded in this picture.[27] The 8×3 unit rectangle (at the upper-left side) gives further evidence to the possibility that the artist was aware of the relationship of the Fibonacci numbers to the golden section. In Seurat's *Bathers at Asnières* (1883, Tate Gallery, London), the golden ratio can be seen as well as several golden rectangles (figure 7-43).

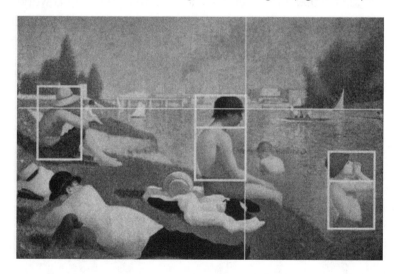

Figure 7-43

Jo Niemeyer (b. 1946), the German graphic artist and architect mentioned earlier, created works according to plans that were based on the golden section. He even calls his works *Geometric Compositions*. In figure 7-44 you will see that his work titled *Utsjoki* displays several golden sections.

27. H. Walser, *Der Goldene Schnitt* (Leipzig: EAG.LE, 2004), p. 135.

Figure 7-44
(Used with permission of Jo Niemeyer.)

For the painting *Variation VI*, 1987 (figure 7-45), we have clear evidence that Niemeyer does not intuitively create his works, but rather he sketches them from a mathematical basis. Notice that he even marked the right side of the sketch (figure 7-45) with the golden ratio (.61803), should our visual assessment of his sketch not suffice.

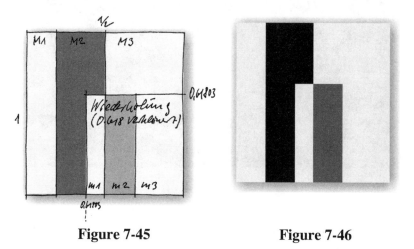

Figure 7-45   **Figure 7-46**

(Used with permission of Jo Niemeyer.)

Niemeyer does not restrict himself to only two-dimensional art; he also works in the third dimension. Again with the sketch of his *Modulon* (figure 7-47) we are given clear evidence of the mathematical underpinnings of his work. *Modulon* is an art object (a cube) divided into sixteen building blocks according to the golden ratio (see figure 7-48).

The possible variations that can be achieved by placing, ordering, grouping, and filling spaces are practically unlimited. The coloring is restricted, however, to the primary colors: blue, red, and yellow, as well as black and white. The *Modulon* has been in the collection of the Museum of Modern Art in New York since 1984.

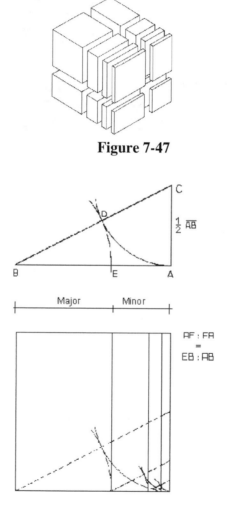

Figure 7-47

Figure 7-48

The Icelandic artist Hreinn Fridfinnsson (b. 1943) clearly based his art on the golden rectangle and the golden spiral. His work called *Untitled* (1988, Kunstmuseum Liechtenstein, Vaduz) can be seen in figure 7-49. This is art based on the golden ratio throughout.

Figure 7-49
(Used with permission of Hreinn Fridfinnsson.)

Another example of an artist's conscious effort to use the Fibonacci numbers as the basic structure of her work is that of the German artist Rune Mields (b. 1935). In her *Evolution: Progression and Symmetry III and IV* (figure 7-50), she exhibits her graphics as representing the Fibonacci numbers. The artist stresses in the accompanying text of the catalog that "[i]n an ascending line, a progression of triangles is generated, with the help of the famous mathematical series of the Leonardo Pisano, called Fibonacci, (0, 1, 1, 2, 3, 5, 8, 13, 21 . . .), who developed it at the beginning of 13th century. It was an assumption that the arising form should be respectively subject to the laws of the symmetry."[28]

---

28. Catalog on the occasion of the exhibitions: Rune Mields—SANCTA RATIO, ed. Kunstverein Lingen Kunsthalle [April 3–May 29, 2005, Kunstverein Lingen, Germany, Kunsthalle, Lingen (Ems, Germany); August 17–Novem: Buxus Verlag, 2005), p. 45.

Figure 7-50
(Used with permission of Rune Mields.)

There is some evidence that the golden section has been embedded in the composition of the following works of art. Whether it was intentional or intuitive is for you to decide. Many of these works can be found in books, on the Internet, or for the world traveler: in person!

- The relief *Dionysius' Procession* (Villa Albani, Rome)
- The fresco *St. Francis Preaching to the Birds* by Giotto (1266–1337)[29] —(Basilica San Francesco, Assisi, Italy)
- The fresco *Trinity* by Masaccio (1401–1429)[30]—(Santa Maria Novella, Florence, Italy)

---

29. Actually Giotto di Bondone, Italian painter and master builder.
30. Actually Tommaso di Giovanni di Simone Gujdi, Italian painter who is regarded as a founder of Renaissance painting. The Trinity fresco advanced the art of mural painting and the development of the altar because of the central perspective.

- The *Deposition from the Cross* by Rogier van der Weyden (ca. 1400–1464)[31] —(altar piece, Prado, Madrid, Spain)
- *The Baptism of Christ* by Piero della Francesca (1415/1420–1492)[32]—(London, National Gallery)
- *Madonna and Child* by Pietro Perugino (ca. 1455–1523)[33] —(Vatican Museum, Rome)
- The painting *The Girl with the Ermine* by Leonardo da Vinci (National Museum, Krakow, Poland)
- The wall painting *The Last Supper* by Leonardo da Vinci (Refectory of Santa Maria delle Grazie, Milan, Italy)
- The round painting of *Madonna Doni* by Michelangelo (1475–1564)[34] —(Uffizi Gallery, Florence, Italy)
- The painting *Crucifixion* (with the Virgin Mary, Saint Jerome, Mary Magdalene, and John the Baptist) by Raphael (National Gallery, London, England)
- The panels *Adam and Eve* by Albrecht Dürer (Prado Museum, Madrid, Spain)
- The painting *The School of Athens* by Raphael (Vatican Museum, Rome)
- The fresco *The Triumph of Galatea* by Raphael (Villa Farnesina, Rome, Italy)
- The copper engraving *Adam and Eve* by Marcantonio Raimondi[35]
- The painting *A Self-portrait* by Rembrandt Harmensz van Rijn (1606–1669)—(National Gallery, London, England)
- *Gelmeroda* by Lyonel Feininger (1871–1956)—(*Gelmeroda* VIII, Whitney Museum, New York; and *Gelmeroda* XII, Metropolitan Museum of Art, New York)
- *Half a Giant Cup Suspended with an Inexplicable Appendage Five Meters Long* by Salvador Dalí (1904–1989)

---

31. Also called Roger de Le Pasture, Flemish painter.
32. Actually Pietro di Benedetto dei Franceschi, also Pietro Borgliese, Italian painter.
33. Actually Pietro Vannucci.
34. Actually Michelangelo Buonarroti; Michelangelo was, however, sculptor, painter, architect, and poet. In 1508 Michelangelo also designed the uniforms of the Swiss Guard for the Vatican.
35. Italian copperplate engraver (ca. 1480–ca. 1534).

The following is a partial list of artists who are often mentioned in connection with the golden section:

- Paul Signac (1863–1935)
- Paul Sérusier (1864–1927)
- Piet Mondrian (1872–1944)[36]
- Juan Gris (1887–1927)[37]
- Otto Pankok (1893–1966)[38]

Paul Sérusier not only knew about the golden section, but he also indicated it in his sketches. In contrast, Gris, Mondrian, and Pankok have flatly denied using it.

Many admirers of ϕ and Fibonacci would be disappointed to learn that the careful examination by the art historian Marguerite Neveux at the end of the twentieth century removed many pictures from the "golden section list." She analyzed x-ray pictures of various canvases and came to the conclusion that most of the artists divided their canvases into eighths before starting their work. There are many ways to use these fractional partitions; yet more often than not, $\frac{5}{8}$ is selected. So if this art historian eliminated some art from the "golden section list," we can still claim them to the "almost golden section" since they used two Fibonacci numbers that can generate a rough approximation of the golden section.

It is truly fascinating how this "magical" golden ratio has inspired artists over the centuries—sometimes deliberately and at other times intuitively. Whether the use of the golden section was conscious or not, no one can deny its prevalence in many of the great masterpieces of the Western world.

---

36. *Painting I* and *Composition with Colored Areas and Gray Lines 1* (1918), for example, are often quoted, as well as *Composition with Gray and Light Brown* (1918; Museum of Fine Arts, Houston, Texas) and *Composition with Red Yellow Blue* (1928).
37. Actually José Victoriano González Pérez, Spanish painter and graphic artist.
38. German painter, graphic artist, and wood engraver.

Chapter 8

The Fibonacci Numbers and Musical Form

Fibonacci Data on the Internet

C an we assume that since you are reading this book, you have already done some research on the Internet about the Fibonacci sequence and its relationship to music?[1] Some of the Web sites one encounters are most fascinating, but what you generally find is often confusing or insipid. Unfortunately, most of the information that shows up on your screen is not meaningful or even accurate. This may be the result of some well-intentioned authors who were writing for elementary school kids and tried to spoon-feed them something about Fibonacci and music that is easily digestible. Well, this stuff now shows up all over the place, and it does not tell you anything generally significant about the relationship. While it is true that the violin is a beautiful example of the application of the golden section to eighteenth-century Italian instrument making, the discus-

---

This chapter was contributed by Dr. Stephen Jablonsky, who is a professor of music and chair of the Music Department at the City College of the City University of New York, as well as a composer.

1. You may have even read some of the silly coincidences that "music" begins with the thirteenth letter of the alphabet, or that in the Dewey Decimal System for library classification, the number for music is 780, which is $2 \cdot 2 \cdot 3 \cdot 5 \cdot 13$—all Fibonacci numbers! (See footnote 16 to Melvil Dewey on p. 211, in chapter 6.)

sions about eight-note scales are completely erroneous. The diatonic scales we all know and love have seven notes, not eight, because number 8 is a repeat of number 1 and continues the scalar motion into the next octave. The fact that the black notes on the piano are divided into twos and threes has nothing to do with the Fibonacci sequence, since it is impossible to tell which one comes first, two or three, because they relate to nothing except the white notes that separate them. A true Fibonacci keyboard would have 2, 3, 5, and 8 black notes in every octave. Now that would be something!

It is also misleading to suggest that certain pleasing frequency ratios, such as 5:3 (a major sixth) and 8:5 (a minor sixth), are related to anything like the Fibonacci sequence because they are only a few among a multitude of ratios that contribute to the rich intervallic (harmonic) essence of the music we know and love. What about the major third (5:4) and minor third (6:5) that are the triadic foundation of all the popular music you have stored in your iPod? And last, if you listen to the pieces that are generated by Fibonacci applications to pitch and/or rhythm, you get to hear music that is perhaps interesting on first hearing, but not worth a second play. So let us talk about Fibonacci applications that really matter.

The golden section has been successfully employed by composers mainly in two areas of the compositional process. The first relates to the location of the climax and the second relates to form. Let us talk about the climax first because that is an easier concept with which to deal, especially if you were not forced to take piano lessons.

The Chopin Preludes

One of the great collections of nineteenth-century piano pieces is the *Preludes* by Frederic Chopin (1810–1849). His book contains twenty-four of the most extraordinary musical miniatures, each one a world unto itself. The first of these is based on an interesting game that Chopin is playing with himself. In figure 8-1 we see the essential melodic activity of the right hand, which is simplified for purposes of illustration. Each measure (except the final six) contains two notes, a step apart, one of which harmonizes with the left hand accompaniment (the whole note) and one that does not (the black note). This piece, which lasts only about a half a minute, is constructed of two dramatic arches of differing sizes. The melody begins on the notes G-A and rises to the E-D in measure 5, where it remains for three measures and then descends back to the G-A in measure 9. From here the melody rises even farther and climaxes on the D-C in measure 21. It then begins its descent to the G-A in measure 25 at which point it leaps twice to E-D before coming to rest on five Cs with A-G pairs below them. The wonderful climax of this miniature masterpiece comes exactly in measure 21 at the golden section of its 34 measures. (Remember these numbers? Yes, they are true members of the Fibonacci sequence! Also recall that $34 \cdot .618 \approx 21$.)

The same accurate placement of the climax at the golden section occurs in *Prelude No. 9 in E major*. This piece is twelve measures long and contains forty-eight beats. The climax occurs exactly on beat 29 ($48 \cdot .618 \approx 29$) at the beginning of the eighth measure. Sometimes it happens and sometimes it doesn't. In any case, the exact location of the climax does not always have to conform to a prescribed mathematical formula. Still, in a large number of cases it occurs in close proximity to the golden mean. Most of the *Preludes* are not in a golden proportion, and they work perfectly well in their less-than-perfect condition. Apparently, Chopin found that it was not necessary to use ϕ to guarantee musical success.

Figure 8-1
Chopin: *Prelude No. 1 in C major*.

Figure 8-2 is the graphic equivalent of figure 8-1 and shows the placement of the golden climax in the sequence of pitches. It was designed for those of you who are not familiar with musical notation.

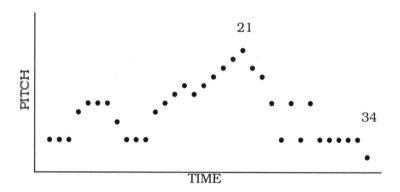

Figure 8-2
Chopin: *Prelude No. 1 in C major.*

Binary Form

Now let us talk about musical form. This may prove to be more taxing, but rest assured, it will be worth it.

Music whose structure contains two parts is said to be in binary form. There are two kinds of binary form: equal binary form is divisible into two parts of relatively equal size, while unequal binary form is divided into two parts of which the second is significantly larger than the first. For a very long time composers have been perfectly happy with the results of using equal binary form—both halves seem to balance each other very nicely. When they use unequal binary form, which may be equally successful, the question arises as to the relative proportions of both parts, and this is where the golden section enters the equation.

We often think of great composers as highly inspired creative types who sit alone in garrets by candle light drawing their inspiration from a distant muse. We imagine their creations to be the result of some reverie when, in fact, most of what they do tends to be quite rational and deliberate. While it is true that there are moments in the creative process when they listen to an inner voice that tells them to do something out of the ordinary, most of the time they are drawing upon years of training and experience that has fostered a technical craft that allows them to make music that "works." By "works" we mean that it conforms to a general game plan that is self-limiting and keeps the piece from becoming a hodgepodge of this, that, and the other thing. The difficult job of the composer is to control the elements of the piece so that it maintains an integral logic but still sounds like it was just made up on the spur of the moment. Most important, the music must present a smooth, continuous narrative that entertains its audience. With very few exceptions, most composers are very bright people who have worked hard to hone their skills and refine their craft. When they put pen to paper (or make music by moving a mouse), they are making very deliberate choices about the sequence of sounds and silences that tells their musical tale. Quite often, the process of composition takes on gamelike qualities. Some of the rules of this game are derived from its style and some are unique to this specific piece. Composers are not only tunesmiths and harmonists; they are often game players.

The Mozart Piano Sonatas

One of the interesting games that have been played by a number of famous composers relates to the use of the golden section in determining the form of a piece. Wolfgang Amadeus Mozart (1756–1791), who loved numbers and all kinds of games, was especially fond of this application. Apparently, when he sat down to write his piano sonatas, he always had the same plan in mind—to try to use the proportions of the golden section to create formal elegance and balance. In Mozart's time, the solo keyboard sonata was comprised of three movements, or separate pieces. The first was spirited and energetic; the second was slow and lyrical; and the third was the fastest of all and brought the sonata to an exciting conclusion.

Before we go any further, be warned that music theory can be a fascinating line of study for those who wish to spend a lifetime in pursuit of truth and beauty, since both can be tantalizingly elusive when it comes to understanding the likes of Mozart. Even the most adept theorists never feel completely satisfied when trying to discover and report on the internal workings of creations like Mozart sonatas. We aim for complete understanding and always settle for something less. Music theory is as complicated as physics and mathematics, but it also has the aesthetic element that seems to take it into the realm of the mystical or magical. Quite often, Mozart seems more the magician than the musician. Such is the case when we study the form of the first movements of his piano sonatas.

One of the most significant contributions to the music of the late eighteenth century was the development of what we call the sonata-allegro form. This name is derived from the fact that it was used almost exclusively in the first movements of all major instrumental forms, all of which may be considered a type of sonata.[2] A symphony is essentially a sonata for an orchestra while a string quartet is a sonata for two violins, a viola, and a cello, and a concerto is a sonata for a soloist and an orchestra. As used in the sonatas, it is a form that has two basic parts, each of which is repeated. The first part is the exposition where the musical materials of the piece are presented. The section is repeated so you can hear all that stuff over again since you probably did not catch it all the first time because it went by so fast. When the repeat is accomplished, we move to the second part that contains the development and the recapitulation. The development does what it says—it is where the materials of the exposition are distorted, chopped up, and tossed about. It is often quite tumultuous and exciting. It is here that the increasing level of tension brings us to the climax of the piece. As the tempest subsides, we return to the beginning of

2. A composition for one or more solo instruments, one of which is usually a keyboard instrument, usually consisting of three or four independent movements varying in key, mood, and tempo.

the piece. That is what recapitulation means—a return to the head. This is where a lot of the sleight of hand takes place, because if we are not paying close attention, we think we are hearing a repeat of the exposition, but we are not. Much has been changed, but it is often quite subtle and goes unnoticed.

Now you have learned more about sonata-allegro form than you wanted to know, but a little extra knowledge won't hurt you. And the next time you hear a symphony or a sonata you will have some idea of how the first movement is put together. As we have seen, the first movement is in binary form (two parts) although the parts are not of equal size. That is the significant factor. The question for Mozart and his contemporaries was how to get these two parts in balance even though they are not the same size. Enter the Fibonacci sequence and the golden mean whose proportions of $\frac{.6180339}{.3819661}$ provide a possible solution. [3]

It should be pointed out that in the baroque period (1600–1750) a great deal of music, especially dance forms, was in equal binary form—both halves were approximately the same size, contained similar music, and were usually repeated. Along come the composers of the classical period (1750–1825), and they expanded this form into what we now know as the sonata-allegro form. This is what we call a rounded binary form because it features a return of the opening section in the latter part of the second. It looks something like this: $\|: A :\|: A^1 \; A :\|$

Mozart wrote eighteen piano sonatas and all but one employ sonata-allegro form in the first movement (the remaining one uses theme and variations form). As may be seen in figure 8-3, of the seventeen, six (35 percent) are exactly divisible by the golden section and are identified as such in the Measures column by the word "golden." Eight (47 percent) are very close and are identified in the Measures column by numbers ranging from −3 to +4. These quantities represent the displacement of the golden section. And three

---

3. This fraction is built by the quotient $\frac{1}{\phi} = 0.618033988 \ldots$ and $\frac{F_{15}}{F_{17}} = \frac{610}{1597} = 0.3819661865 \ldots$.

(18 percent) are really not close enough for serious consideration because the expositions are six, eight, and twelve measures too long. Statistically, that certainly leaves one with the impression that the use of the golden section was important to Mozart.

Mozart Sonata	Key	Length	Exposition	Proportion	Accuracy
No. 1, K. 279	C major	100	38	0.38	golden
No. 2, K. 280	F major	144	56	0.389	golden
No. 3, K. 281	Bb major	109	40	0.367	-2
No. 4, K. 282	Eb major	36	15	0.417	1
No. 5, K. 283	G major	120	53	0.442	8
No. 6, K. 284	D major	127	51	0.402	3
No. 7, K. 309	C major	156	59	0.378	golden
No. 8, K. 310	A minor	133	49	0.368	-1
No. 9, K. 311	D major	112	39	0.348	-3
No. 10, K. 330	C major	149	57	0.383	golden
No. 11, K. 331	A major	135	55	Theme & Var.	
No. 12, K. 332	F major	229	93	0.406	6
No. 13, K.333	Bb major	170	63	0.371	-1
No. 14, K.457	C minor	185	74	0.4	4
No. 15, K.545	C major	73	28	0.384	golden
No. 16, K.570	Bb major	209	79	0.378	golden
No. 17, K. 576	D major	160	58	0.363	-2
No. 18, K. 533	F major	240	103	0.429	12

**Figure 8-3
Mozart Piano Sonatas.**

Of the six that are right on the money, there is one, Sonata No. 1 (K.279), that is exactly 100 measures in length and the exposition ends in measure 38. It just does not get more obvious than that, and because it is the first, it serves as a kind of declaration of purpose. Here is what it looks like:

Exposition   :||: Development   Recapitulation :||

38 MEASURES   62 MEASURES

Figure 8-4

Those of you who know something about the music of the late eighteenth century might be asking yourself whether Franz Joseph Haydn (1732–1809) was just as keen to use the golden section in his piano sonatas. After all, he is the other giant of the classical period and it would be interesting to see his approach to sonata-allegro form. As it turns out, his adherence to ϕ did not match Mozart's. In analyzing an equal number of randomly chosen piano sonatas, we find that only 18 percent $\left(\frac{3}{17}\right)$ of his sonata-allegro forms are golden, while 53 percent $\left(\frac{9}{17}\right)$ are close, and 29 percent $\left(\frac{5}{17}\right)$ are out of consideration. Make of it what you will.

Haydn Sonata	Key	Length	Exposition	Proportion	Accuracy
No. 14, 1767	E major	84	30	0.357	-2
No. 15, 1767	D major	110	36	0.327	-6
No. 16, 1767	Bb major	116	38	0.327	-6
No. 17, 1767	D major	103	42	0.408	2
No. 19, 1773	C major	150	57	0.38	golden
No. 21, 1773	F major	127	46	0.362	-3
No. 25, 1776	G major	143	57	0.399	2
No. 26, 1776	Eb major	141	52	0.369	-2
No. 27, 1776	F major	90	31	0.344	-4
No. 31, 1778	D major	195	69	0.353	-6
No. 32, 1778	E minor	127	45	0.354	-4
No. 33, 1780	C major	172	68	0.395	2
No. 34, 1780	C# minor	100	33	0.33	-5
No. 35, 1780	D major	103	40	0.388	golden
No. 42, 1786	G minor	77	30	0.39	golden
No. 43, 1786	Ab major	112	38	0.339	-5
No. 49, 1793	Eb major	116	43	0.371	-1

Figure 8-5
Haydn Piano Sonatas.

So, are the Mozart movements better than those of Haydn? Can a statistical analysis help us arrive at a reasonable conclusion? The proportions of the average Mozart movement are $\frac{.389}{.611}$ as compared to Haydn's $\frac{.364}{.636}$. However, Mozart's placement of the double bar relative to ϕ ranged from –3 to +12, while Haydn's only ranged from –6 to +2. Perhaps there are other ways of using the data to help us. But it is also possible that numbers may have nothing to do with the quality of the music. Here is a suggestion: listen to all of the thirty-four movements under consideration in pairs—one Mozart followed by one Haydn—and see how it goes. Even though you may be as confused at the end of the test as you were at the beginning, at least you will have heard a lot of terrific music.

Beethoven's Fifth Symphony

The opening five measures of the first movement of this symphony forms probably the most universally recognized musical statement in classical music. Few of the millions who adore this piece realize how extraordinary and revolutionary it is in the history of Western classical music. There are so many aspects to this work that are worthy of discussion that countless authors have written tomes about it. Since this is a book about the Fibonacci sequence, let us try to contain our enthusiasm and stick to the topic (it may not be easy).

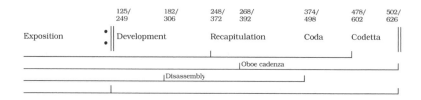

Figure 8-6
Beethoven, *Symphony No. 5, first movement.*

Before you get dazzled by the data, you need to understand how the form of this movement differs from those in the sonatas and symphonies of Mozart and Haydn. Ludwig van Beethoven (1770–1827) was born into the generation of musicians that followed these two pillars of Austro-Hungarian musical culture and proved to be the most revolutionary and influential composer of the Romantic period. Surprisingly, the exposition, development, and recapitulation in this astounding work are all essentially the same size, thus preventing a golden section of these three sections. Instead of bringing the first movement to a conclusion, the recapitulation rushes headlong into a coda[4] that turns out to be another development section—something his predecessors never contemplated. Compounding that is the addition of a codetta (little tail section) to the coda which, when taken together, add a significant and unprecedented fifth section. This tail is big enough to wag the dog! Therefore, we end up with five sections, not four, all of which are between 124 and 128 measures. This is a new kind of sonata-allegro form.

As may be seen in figure 8-6, this newly expanded sonata-allegro form contains three golden sections that may have been part of a deliberate attempt to apply the principle in new ways. First, the arrival of the recapitulation in measure 372 is the golden section of the entire piece without the shocking second coda. Without the codetta, the piece would be 602 measures in length: $602 \cdot .618 \approx 372$. Right on the mark! The rest of the evidence hits the target—not exactly in the bull's eye, but close enough for music.

The end of the repeat of the exposition (measure $124 \cdot 2 =$ measure 248) is almost a golden section of the entire movement $\left(\frac{248}{626} \approx .396 \right)$. If we remove the two measures of rest at the end of the exposition (measures 123–24) the proportion is even closer $\left(\frac{244}{626} \approx .389 \right)$. In a generous mood we could credit Beethoven for that one too.

Two very special compositional events occur at points of critical proportion. The first occurs in the development, where the four-note motive begins to break into two notes and then into one. This disassembly of the four-note motive, that has remained intact so far, occurs at measure 306. This is a golden section of the movement up to the end of the recapitulation in measure 498 $\left(\frac{306}{498} \approx .614 \right)$. One of the most inspired moments in the entire symphony occurs in the recapitulation when, in measure 392, the entire orchestra stops, as it did in the analogous spot in the exposition, except for the oboe. The lit-

---

4. A coda is a section of music added at the end of a composition, introduced to bring it to a satisfactory close.

tle cadenza, which the oboe plays, has no precedent and is dramatically shocking and perplexing to those familiar with sonata-allegro form. This astounding solo occurs only six measures from the golden section of the entire movement (626 · .618 = measure 386). "Close, but no cigar," you say. Well, we will never know if Beethoven had intended this special moment to occur at a golden section, but it is tantalizingly close, and it is possible that, in the heat of creation, measures were added and subtracted with the final result as we see it now. The process of continually reediting his music was Beethoven's usual practice, for he wrestled mightily until his almost indecipherable scores were finally finished and could be sent to befuddled copyists and publishers.

Wagner's Prelude to *Tristan und Isolde*

When, in the late 1850s, Richard Wagner (1813–1883) sat down to write a mythic opera about the delays and disappointments of tragic love, he had a plan to force a gigantic leap in the art of musical composition. So revolutionary is this work that today theorists are still arguing about its construction. We will bypass all that extraordinary stuff, however, and get to the ϕ at the heart of the prelude that opens this masterpiece.

Wagner, like Mozart, was a composer who liked to play games. Perhaps the most extraordinary evidence of Wagner's game playing is his deliberate use of the golden section to delineate significant keys in the *Tristan* prelude. By "key" we mean what scale, or collection of notes, is being used as the primary material of a piece of music. C major is a scale that contains the notes C, D, E, F, G, A, B, and C. Beginning pianists like this scale because it is comprised of only white notes, which are easier to play. Anyway, a piece that is in C major uses the notes of that scale as the building blocks of its composition. At the beginning of every line of music there is what we call a key signature. This is a collection of flats or sharps that composers put there so they do not have to notate the flats and sharps next to every note. It is, in effect, a form of musical short hand. For example, if you see a sharp (#) on the F line of the staff in the key signature, it means that the performer should play only F sharps (black notes), not F naturals (white notes).

It should be mentioned that European intellectuals in the 1850s were familiar with the principle of the golden section, since it had come back into vogue around this time. Wagner certainly knew about it, but would not have let slip that he was using this principle. He, like so many composers, did not like to tell his audience how he did his tricks of musical magic. These composers assumed that trained musicians who were curious would dig into their scores in order to unearth the hidden secrets. Until recently, most Wagner theorists did not realize that the golden section applications were at work in this piece. They lay undiscovered for more than 140 years.

An examination of the score of the prelude reveals something very strange. The key signature of the piece has no flats or sharps, indicating that the scale being used is either C major or A minor. In this case, it is A minor. In measure 43, the key signature changes to A major (three sharps) even though the key is not A major. In measure 71 the key signature returns to A minor. What is strange about this is the fact that the beautiful, but revolutionary, music you hear is never actually in any key—the tonal orientation moves frequently from key to key. So the question is, Why would Wagner go to the trouble of changing key signatures twice in a piece that really isn't in any key? The extreme fluctuation of keys does not seem to call for a key signature, and both changes take place in the middle of phrases. If you chart this phenomenon, it looks like this:

|--- 42 measures --- || --- 28 measures --- || --- 41 measures ---|

Figure 8-7

Look familiar? Have you done the math? The double bar at the end of measure 70 divides the 111 measures by a section that is extremely close to the golden section $\left(\frac{70}{111}=.\overline{630}\right)$, and the double bar

at the end of measure 42 does the same thing for the first 70 meas-
ures $\left(\frac{42}{70}=.6\right)$. This is only one of the many fascinating ways this
piece is organized in terms of key structure. There are applications
of the golden sections all over the place, as you can see by looking
at the diagram in figure 8-8.

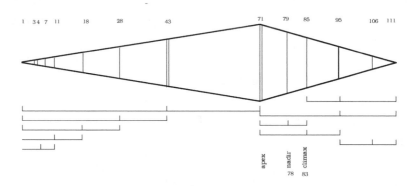

Figure 8-8
Wagner: Prelude to *Tristan und Isolde*.

The harmonic events that take place at each of these points of de-
marcation may not mean much to you unless you are a scholar of
Wagner's chromaticism and shifting tonal centers. But rest assured, these events are critical to the construction of the piece.
Wagner knew what he was doing. If you are wondering where
Wagner put his climax, it is not at the golden section (measure 68),
but in measure 83, three-quarters of the way through the
piece—very late in the game by traditional standards.

The following information is only for musicians, the foolishly
curious, or mathematical sticklers. The golden mean of a piece 111
measures   long   is   calculated   to   be   measure   68.
$\left(\text{obtained by: } 111\cdot.618 \approx 68.6\right)$. The question is: Should we expect
Wagner to do something really special at the golden mean? The an-
swer is: Sure, why not? While it is true that he placed the double
bars at the end of measure 70, measures 68–69 are significant be-

cause they contain the last of the dominants[5] in A minor (E^9) that began in measure 63, and this final dominant is preceded by a supertonic Bø. This is the only true ii–V progression in the tonic key[6] (Am) in the entire piece! The strategic importance of measures 69–71 is reinforced not only by the return to the opening key signature but also by the shift away from E^9 to G^9, the dominant of C major, as well as the arrival at the apex—the only consecutive return of the opening A–F dyad[7] in either the melody or the bass line!

Bartok's *Music for Strings, Percussion, and Celesta*

When, in 1936, Bela Bartok (1881–1945) was commissioned by Paul Sacher (1906–1999) to write a work for chamber orchestra, he decided to provide the wealthy conductor with something very special. What he produced was the *Music for Strings, Percussion, and Celesta*, one of the seminal works of the twentieth century. It is a piece that uses no woodwinds or brass instruments in its ensemble. What we hear is a small string orchestra in combination with a variety of percussion instruments: side drum, snare drum, cymbals, tam-tam, bass drum, tympani, xylophone, celesta, harp, and piano. This instrumentation had never been used before, or has been since. Its sound is totally unique. It is quintessential Bartok.

This masterpiece is comprised of four movements, the first of which is a fugue. Just in case you forgot what you learned in Music 101, a fugue is a contrapuntal (two or more melodies at the same time), additive musical process that begins with a solo statement of the fugue subject (melody). When the statement ends (it can be anywhere from a few notes to a protracted melody), it is imitated in another voice (or instrument) at a different pitch level (higher or lower). At this point you hear the fugue subject in the second voice, accompanied by a countermelody in the continuation of the

---

5. A dominant is the fifth note of a scale, or the chord built on that note.
6. A tonic is the first note of a scale.
7. A dyad is any two notes or pitches (consecutively or simultaneously), also known as an interval, but dyad sounds better when the two pitches in question are important.

first voice. Most fugues have three or four voices, and each of them gets to state the subject in the opening section we call the exposition. When each of the voices has had a chance to present the subject, and we have three or four countermelodies all playing together, we move on to the development section. It is here that just about anything can happen. The subject may be played by one or more voices in inversion (upside down), backward; it may be rhythmically altered or chopped into pieces. There may also be a lot of overlapping of subject statements, something we call *stretto*. There is no prescription for what should be done. That is left to the composer's imagination and invention. It is a form that was quite popular in the baroque period, and Johann Sebastian Bach (1685–1750) was undoubtedly its master. Composers since Bach have tried their hand at this complicated contrapuntal challenge, but fugues were relatively rare in the nineteenth and twentieth centuries, and are so far in the twenty-first century. It is not an easy thing to write because it is the musical equivalent of a crossword puzzle that makes sense both vertically and horizontally at the same time.

Two centuries after Bach, Bartok provides us with one humdinger of a fugue in the first movement of his *Music*. What he did was take an old idea and dress it up in new clothes. It is, in effect, a super fugue for the twentieth century. The most striking update is the presentation of the fugue subject on all twelve steps of the chromatic scale (all the black and white notes within one octave) beginning with A. The arrival on the twelfth pitch (E flat) in measure 56 is also the climax of the piece and is very close to golden in its proportion $\left(\frac{56}{88}=.\overline{63}\right)$. From this point to the end of the piece, the statements work their way back through all twelve pitches until the final statement on A. Figure 8-9 shows how the subject entrances are related by pitch. The first statement begins on A and the second is a fifth higher on E. The third entrance is a fifth lower than A (D) and is followed by an entrance a fifth high than E (B). The pattern of fifths continues until measure 56, where we find all the strings playing in unison E flats. The scheme works backward from there until the end:

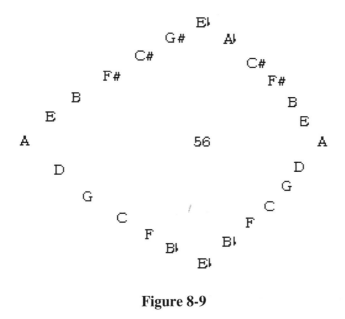

Figure 8-9

If you take a good look at the diagram in figure 8-10, you will notice that it bears an uncanny resemblance to figure 8-8. The major difference is the placement of the climaxes, with Bartok being more of a traditionalist than was Wagner seventy-seven years earlier. This diagram (figure 8-10) displays the significant golden section and its relationship to subject statements or entrances of new instrumental colors. It is obvious from the pitch scheme and the use of the golden section that this piece is one enormously complicated game. There are other games at work in this score, but they require more training and patience on the part of the reader than may be appropriate here. But, lest you feel cheated, one or two more delicious morsels couldn't hurt. The fugue subject is comprised of four little phrases, each of which rises and falls like a miniature version of the grand scheme of the piece. When we come to the last phrase of the fugue, we find the second phrase of the subject played by the first violin, while the second violin plays its inversion. This simultaneous, mirrored bow form (A–E flat–A) looks and sounds like a miniature version of the diagram you see below.

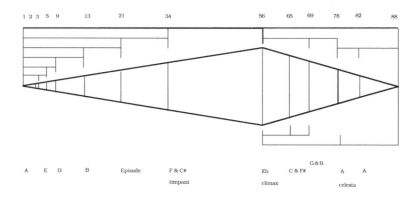

Figure 8-10
Bartok: *Music for Strings, Percussion, and Celesta.*

Good Math, Bad Music

A possible reason that much contemporary music has never been well received may be that too many composers got lost in the number games of twentieth-century compositional practice. While it is true that all composers play mind games with themselves in the process of creating music, many luminaries in the generation that included the likes of Pierre Boulez (b. 1925) and Milton Babbitt (b. 1916) lost sight of the fact that people are mostly emotional beings, not numerical calculators (sacred hearts, not sacred brains). Too many of the works from that period are brilliantly calculated but are devoid of the color, mood, passion, or musical narrative that can move an audience to tears or rapture.

When Arnold Schönberg (1874–1951) applied his twelve-tone method (formulated in 1924), he imbued his soundscapes with the nineteenth-century Romanticism of his youth. He was always the Romantic. Here was a brilliant mind that was searching for a replacement for the diatonic tonal system and, in the process, created exquisitely crafted works that few people wanted to hear. His music had a new technology clothed in an out-of-date aesthetic. It was his student, Anton von Webern (1883–1945), who seemed to apply an aesthetic that was truly modern, more in tune with the times and the new note-picking system. For a long time performers of his music did not realize that he, too, was a Romantic, and their performances reflected that ignorance—they were insipid. He just disguised his romanticism better than Schönberg. His sparse textures must be performed with the same passion that one would apply to the music of Johannes Brahms (1833–1897) and Gustav Mahler (1860–1911), only the attention is focused on far fewer notes.

This misunderstanding of Webern's underlying Romanticism led a lot of young composers that followed him to believe they could play all those wonderful numerical and structural mind games and everything would work out just fine. Well, it did not. Over many decades the musically adventurous among us have attended too many concerts where minds were dazzled, but hearts were unmoved. Often they went home unable to remember a single musical gesture. Too often they heard a vast amount of musical data without hearing anything that seemed to be meaningful. What this music often lacked was modesty. Mozart played as many mind games as anyone who ever lived, only he hid his calculations where only the refined ear and eye could find them. His out-front, tuneful lyricism disguised his underlying machinery. His music invites us to sing along, and, once we are seduced by his magic, we spend a lifetime trying to figure out how he did it.

Music is a communication from heart to heart and mind to mind—it must tell us what it means to be a human being, not a UNIVAC.[8] The most difficult part of being a composer is trying to balance the heart and the mind, and it does not take much to tilt slightly in the wrong direction with results that are brilliantly insipid and vacuous. There are those, like minimalist composer Tom Johnson (b. 1939), who believe otherwise. Here is a fellow who has employed a great deal of mathematical calculations in his compositional process with questionable results. Many art forms in

---

8. UNIVAC is an acronym, standing for UNIVersal Automatic Computer. UNIVAC was one of the eight major computer companies throughout most of the 1960s.

the twentieth century have had practitioners who eschew their own humanity. For them, emotion plays no part in the artistic process. Johnson admits this denial of the personal element as follows:

> I have often tried to explain that my music is a reaction against the romantic and expressionistic musical past, and that I am seeking something more objective, something that doesn't express my emotions, something that doesn't try to manipulate the emotions of the listener either, something outside myself. Sometimes I explain that my reasons for being a minimalist, for wanting to work with a minimum of musical materials, is because it also helps me to minimize arbitrary self-expression. Sometimes I say, "I want to find the music, not to compose it."[9]

This is in tune with the philosophy originally promoted by John Cage (1912–1992), a highly influential thinker whose music has never found wide acceptance with a paying audience. Many who have heard Johnson's music feel that he fulfills his minimalist intentions most effectively. Pieces such as his *Narayana's Cows*, a composition that employs the Fibonacci series as a pitch generator, are pleasing to some, but have not found wide acceptance. Many people, when listening to his music, come away disappointed that they have not been touched by the personal revelations of a sentient being. They experience the musical equivalent of eating a vegetarian snack—while at the concert they are entertained by a skillful array of sounds, but an hour later they are hungry for something more profound.

It's too often the case that good math doesn't make good music.

The Coda

If you are still reading, it means that you are of stout heart and firm mind because, for a book that was supposed to be about math, you

---

9. http://www.ChronicleoftheNonPopRevolution,Kalvos.org/johness4.html.

have been inundated by an awful lot of music theory. But, if you got this far, you can see that though the golden mean plays a significant role in music, it is by no means the whole story. Below you will find a little reminder of Fibonacci in music: a picture of a violin with its proportions related to ϕ. It is, unquestionably, a lovely shape to behold and, when put in the right hands, can be an extraordinary conveyor of passionate expression. Antonio Stradivarius (1644–1737) is, undoubtedly, the most famous maker of violins. His instruments set a standard that is still being followed today. The proportions, components, and assembly of his instruments have been studied and imitated by all those attempting to replicate an instrument that sings freely and projects to the farthest reaches of the concert hall. Today, his instruments cost several million dollars on those rare occasions when they come on the market. If you can't make the payments on a "Strad" and want to go out to the shop to see what you can paste together, here are the Fibonacci proportions to get you started. Good luck!

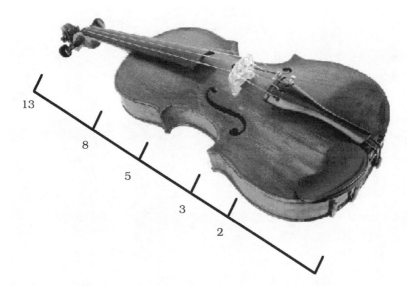

Figure 8-11

Chapter 9

The Famous Binet Formula for Finding a Particular Fibonacci Number

Until now, we usually inspected the Fibonacci numbers by their location in their sequence. If we wanted to find the tenth Fibonacci number, we would write the sequence to 10 numbers, and then we would have the tenth number. In other words, we would simply list them as 1, 1, 2, 3, 5, 8, 13, 21, 34, 55 and count to the tenth number in the sequence. This would give us 55 as the tenth Fibonacci number. Such a procedure can be somewhat cumbersome, however, if we seek, say, the fiftieth Fibonacci number. After all, listing fifty Fibonacci numbers is no easy matter if you don't have the list handy that we offer you in appendix A.

Well, the French mathematician Jacques-Philippe-Marie Binet (1786–1856) provided us with a formula for finding any Fibonacci number without actually listing the sequence as we just did. To develop this formula, we will gently take you through some elementary algebra, explaining all of our steps along the way. But before we embark on this somewhat tedious task, we need to understand what it means to derive such a formula. We will, thus, examine a simple and very well-known kind of number—the square number—to determine what it means to work with a recursive sequence of numbers such as the Fibonacci numbers.

Figure 9-1
Jacques-Philippe-Marie Binet.

The recursive nature of the Fibonacci numbers comes from the definitive relationship: $F_{n+2} = F_n + F_{n+1}$, where we know that $F_1 = 1$ and $F_2 = 1$. We are accustomed to finding any Fibonacci number in the sequence as long as we know its two preceding numbers.

Suppose we seek to find the sixth square number (remember, the square numbers are 1, 4, 9, 16, 25, 36, 49, 64, 81, and so on), then we simply square 6 to get 36. If we are called on to find the 118th square number, then we square 118 to get $118 \cdot 118 = 13,924$. This is an explicit definition of the square numbers. Let's try to set up a recursive definition of the square numbers.

We begin with the definition of S_n, which will represent the nth square number. We know that $S_1 = 1$, and we will see that $S_{n+1} = S_n + (2n + 1)$. Using this definition, we would get the following:[1]

---

1. We can also use that $S_0 = 0$, the 0th square number.

$$S_1 = S_0 + (2 \cdot 0 + 1) = 0 + 1 = 1$$
$$S_2 = S_1 + (2 \cdot 1 + 1) = 1 + 3 = 4$$
$$S_3 = S_2 + (2 \cdot 2 + 1) = 4 + 5 = 9$$
$$S_4 = S_3 + (2 \cdot 3 + 1) = 9 + 7 = 16$$
$$S_5 = S_4 + (2 \cdot 4 + 1) = 16 + 9 = 25$$

This tells us that we can get a square number by adding to its predecessor-square number a number that is one greater than twice the position number of that square. Of course, you may be thinking that this is a silly way to look at a rather simple problem—when you can just square the number and that's it. However, we are trying to show that there is a recursive relationship from which one can also generate square numbers—even though it is ridiculously more complicated.

From this we can also conclude that the sum of consecutive odd numbers (beginning with 1) is always a square number, which leads us to the recursive relationship as displayed below.

Since $1 + 3 + 5 + \ldots + (2n - 1) + (2n + 1) = (n + 1)^2$, and $(n + 1)^2 = n^2 + (2n + 1)$, for any natural number n, we get the recursive relationship we used above.

Even though we can easily prove this by the technique known as mathematical induction, we can see this in a geometric representation (figure 9-2), which can serve a similar purpose.

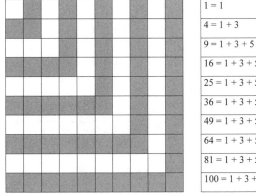

$1 = 1$
$4 = 1 + 3$
$9 = 1 + 3 + 5$
$16 = 1 + 3 + 5 + 7$
$25 = 1 + 3 + 5 + 7 + 9$
$36 = 1 + 3 + 5 + 7 + 9 + 11$
$49 = 1 + 3 + 5 + 7 + 9 + 11 + 13$
$64 = 1 + 3 + 5 + 7 + 9 + 11 + 13 + 15$
$81 = 1 + 3 + 5 + 7 + 9 + 11 + 13 + 15 + 17$
$100 = 1 + 3 + 5 + 7 + 9 + 11 + 13 + 15 + 17 + 19$

Figure 9-2

We have now seen the square numbers from the explicit definition (our more common one) and the recursive definition (the much more cumbersome one). Until now, we have been viewing the Fibonacci numbers from the recursive relationship. In 1843 Jacques-Philippe-Marie Binet (1786–1856)[2] developed an explicit definition for the Fibonacci numbers. As is often the case in mathematics when a formula is named after a mathematician, controversies arise as to who was actually the first to discover it. Even today when a mathematician comes up with what appears to be a new idea, others are usually hesitant to attribute the work to that person. They often say something like: "It looks original but how do we know it wasn't done by someone else earlier?" Such is the case with the Binet formula. When he publicized his work, there were no challenges to Binet, but in the course of time, there have been claims that Abraham de Moivre (1667–1754) was aware of it in 1718, Nicolaus Bernoulli I (1687–1759) knew it in 1728, and his cousin Daniel Bernoulli (1700–1782)[3] also seems to have known the formula before Binet. The prolific mathematician Leonhard Euler (1707–1783) is also said to have known it in 1765. Nevertheless, it is known today as the *Binet formula*.

Let us now slowly try to develop this relationship. We will begin by recalling the golden ratio $\left(\frac{1}{x}=\frac{x}{x+1}\right)$ that led us to the equation:

$x^2 - x - 1 = 0$. The roots of this equation are ϕ and $-\frac{1}{\phi}$, where $\phi = \frac{\sqrt{5}+1}{2}$ is the announced golden ratio and therefore is $\frac{1}{\phi} = \frac{2}{\sqrt{5}+1} = \frac{\sqrt{5}-1}{2}$.

By now you may be familiar with the sum and difference of these terms as:

$$\phi + \frac{1}{\phi} = \frac{\sqrt{5}+1}{2} + \frac{\sqrt{5}-1}{2} = \sqrt{5} \quad \text{and} \quad \phi - \frac{1}{\phi} = \frac{\sqrt{5}+1}{2} - \frac{\sqrt{5}-1}{2} = 1$$

---

2. " Memoire sur l'integration des equations lineaires aux diffeences finies d'un ordre quelconque, a coefficients variables," *Comptes rendus de l'academie des sciences de Paris*, vol. 17, 1843, p. 563.
3. The Bernoulli family is like a clan (of 8 mathematicians in 3 generations)—famous and estranged!

As a next step, we will take successive powers of this sum and difference. Figure 9-3 shows this.

n	Sum		Difference	
1	$\phi + \dfrac{1}{\phi}$	$= \mathbf{1}\sqrt{5}$	$\phi - \dfrac{1}{\phi}$	$= 1$
2	$\phi^2 + \dfrac{1}{\phi^2}$	$= 3$	$\phi^2 - \dfrac{1}{\phi^2}$	$= \mathbf{1}\sqrt{5}$
3	$\phi^3 + \dfrac{1}{\phi^3}$	$= \mathbf{2}\sqrt{5}$	$\phi^3 - \dfrac{1}{\phi^3}$	$= 4$
4	$\phi^4 + \dfrac{1}{\phi^4}$	$= 7$	$\phi^4 - \dfrac{1}{\phi^4}$	$= \mathbf{3}\sqrt{5}$
5	$\phi^5 + \dfrac{1}{\phi^5}$	$= \mathbf{5}\sqrt{5}$	$\phi^5 - \dfrac{1}{\phi^5}$	$= 11$
6	$\phi^6 + \dfrac{1}{\phi^6}$	$= 18$	$\phi^6 - \dfrac{1}{\phi^6}$	$= \mathbf{8}\sqrt{5}$
7	$\phi^7 + \dfrac{1}{\phi^7}$	$= \mathbf{13}\sqrt{5}$	$\phi^7 - \dfrac{1}{\phi^7}$	$= 29$
8	$\phi^8 + \dfrac{1}{\phi^8}$	$= 47$	$\phi^8 - \dfrac{1}{\phi^8}$	$= \mathbf{21}\sqrt{5}$
9	$\phi^9 + \dfrac{1}{\phi^9}$	$= \mathbf{34}\sqrt{5}$	$\phi^9 - \dfrac{1}{\phi^9}$	$= 76$
10	$\phi^{10} + \dfrac{1}{\phi^{10}}$	$= 123$	$\phi^{10} - \dfrac{1}{\phi^{10}}$	$= \mathbf{55}\sqrt{5}$

Figure 9-3

An inspection of the values generated for the sum and difference suggests that we can represent them in terms of Fibonacci numbers (F_n) and Lucas numbers (L_n), as in figure 9-4.

n	Sum		Difference	
1	$\phi + \dfrac{1}{\phi}$	$= 1\sqrt{5} = F_1\sqrt{5}$	$\phi - \dfrac{1}{\phi}$	$= 1 = L_1$
2	$\phi^2 + \dfrac{1}{\phi^2}$	$= 3 = L_2$	$\phi^2 - \dfrac{1}{\phi^2}$	$= 1\sqrt{5} = F_2\sqrt{5}$
3	$\phi^3 + \dfrac{1}{\phi^3}$	$= 2\sqrt{5} = F_3\sqrt{5}$	$\phi^3 - \dfrac{1}{\phi^3}$	$= 4 = L_3$
4	$\phi^4 + \dfrac{1}{\phi^4}$	$= 7 = L_4$	$\phi^4 - \dfrac{1}{\phi^4}$	$= 3\sqrt{5} = F_4\sqrt{5}$
5	$\phi^5 + \dfrac{1}{\phi^5}$	$= 5\sqrt{5} = F_5\sqrt{5}$	$\phi^5 - \dfrac{1}{\phi^5}$	$= 11 = L_5$
6	$\phi^6 + \dfrac{1}{\phi^6}$	$= 18 = L_6$	$\phi^6 - \dfrac{1}{\phi^6}$	$= 8\sqrt{5} = F_6\sqrt{5}$
7	$\phi^7 + \dfrac{1}{\phi^7}$	$= 13\sqrt{5} = F_7\sqrt{5}$	$\phi^7 - \dfrac{1}{\phi^7}$	$= 29 = L_7$
8	$\phi^8 + \dfrac{1}{\phi^8}$	$= 47 = L_8$	$\phi^8 - \dfrac{1}{\phi^8}$	$= 21\sqrt{5} = F_8\sqrt{5}$
9	$\phi^9 + \dfrac{1}{\phi^9}$	$= 34\sqrt{5} = F_9\sqrt{5}$	$\phi^9 - \dfrac{1}{\phi^9}$	$= 76 = L_9$
10	$\phi^{10} + \dfrac{1}{\phi^{10}}$	$= 123 = L_{10}$	$\phi^{10} - \dfrac{1}{\phi^{10}}$	$= 55\sqrt{5} = F_{10}\sqrt{5}$

Figure 9-4

Focusing on the Fibonacci numbers, we notice that they appear alternately as coefficients in the sums and differences of the powers of $\phi = \dfrac{\sqrt{5}+1}{2}$ and $\dfrac{1}{\phi} = \dfrac{\sqrt{5}-1}{2}$. For even-number powers they appear as the difference, and for odd-number powers they appear as the sum. This can be handled by using -1 to various powers; since when -1 is taken to an odd power, the result is negative, and when it is taken to an even power, the result is positive. This can be summarized with the following expression:

$$\phi^n - (-1)^n \frac{1}{\phi^n} = \phi^n - \left(-\frac{1}{\phi}\right)^n$$

Remember, in figure 9-4 each of the Fibonacci numbers was multiplied by $\sqrt{5}$, so in order for us to have an expression equal to the Fibonacci numbers, we need to divide our result by $\sqrt{5}$. Therefore,

$$F_n = \frac{1}{\sqrt{5}}\left[\phi^n - \left(-\frac{1}{\phi}\right)^n\right]$$

A summary of this division is seen in figure 9-5.

n	$\phi^n - \left(-\dfrac{1}{\phi}\right)^n$			$\dfrac{1}{\sqrt{5}}\left[\phi^n - \left(-\dfrac{1}{\phi}\right)^n\right]$	F_n
1	$\phi - \left(-\dfrac{1}{\phi}\right)$	$= 1\sqrt{5}$		$\dfrac{1}{\sqrt{5}}\left[\phi - \left(-\dfrac{1}{\phi}\right)\right]$	$= 1$
2	$\phi^2 - \left(-\dfrac{1}{\phi}\right)^2$	$= 1\sqrt{5}$		$\dfrac{1}{\sqrt{5}}\left[\phi^2 - \left(-\dfrac{1}{\phi}\right)^2\right]$	$= 1$
3	$\phi^3 - \left(-\dfrac{1}{\phi}\right)^3$	$= 2\sqrt{5}$		$\dfrac{1}{\sqrt{5}}\left[\phi^3 - \left(-\dfrac{1}{\phi}\right)^3\right]$	$= 2$
4	$\phi^4 - \left(-\dfrac{1}{\phi}\right)^4$	$= 3\sqrt{5}$		$\dfrac{1}{\sqrt{5}}\left[\phi^4 - \left(-\dfrac{1}{\phi}\right)^4\right]$	$= 3$
5	$\phi^5 - \left(-\dfrac{1}{\phi}\right)^5$	$= 5\sqrt{5}$		$\dfrac{1}{\sqrt{5}}\left[\phi^5 - \left(-\dfrac{1}{\phi}\right)^5\right]$	$= 5$
6	$\phi^6 - \left(-\dfrac{1}{\phi}\right)^6$	$= 8\sqrt{5}$		$\dfrac{1}{\sqrt{5}}\left[\phi^6 - \left(-\dfrac{1}{\phi}\right)^6\right]$	$= 8$
7	$\phi^7 - \left(-\dfrac{1}{\phi}\right)^7$	$= 13\sqrt{5}$		$\dfrac{1}{\sqrt{5}}\left[\phi^7 - \left(-\dfrac{1}{\phi}\right)^7\right]$	$= 13$
8	$\phi^8 - \left(-\dfrac{1}{\phi}\right)^8$	$= 21\sqrt{5}$		$\dfrac{1}{\sqrt{5}}\left[\phi^8 - \left(-\dfrac{1}{\phi}\right)^8\right]$	$= 21$
9	$\phi^9 - \left(-\dfrac{1}{\phi}\right)^9$	$= 34\sqrt{5}$		$\dfrac{1}{\sqrt{5}}\left[\phi^9 - \left(-\dfrac{1}{\phi}\right)^9\right]$	$= 34$
10	$\phi^{10} - \left(-\dfrac{1}{\phi}\right)^{10}$	$= 55\sqrt{5}$		$\dfrac{1}{\sqrt{5}}\left[\phi^{10} - \left(-\dfrac{1}{\phi}\right)^{10}\right]$	$= 55$

Figure 9-5

The test to see if we, in fact, have developed a formula for generating any Fibonacci number is for us to apply the formula to our now-familiar recursive relationship $F_{n+2} = F_n + F_{n+1}$.

This would require a proof by mathematical induction. (For this we refer you to appendix B.)

By substituting the value of ϕ in $\left[\phi^n - \left(-\dfrac{1}{\phi} \right)^n \right]$, we get the Binet formula in terms of real numbers.

$$F_n = \frac{1}{\sqrt{5}}\left[\phi^n - \left(-\frac{1}{\phi} \right)^n \right] = \frac{1}{\sqrt{5}}\left[\left(\frac{\sqrt{5}+1}{2} \right)^n - \left(-\frac{1}{\frac{\sqrt{5}+1}{2}} \right)^n \right]$$

$$= \frac{1}{\sqrt{5}}\left[\left(\frac{1+\sqrt{5}}{2} \right)^n - \left(\frac{1-\sqrt{5}}{2} \right)^n \right]$$

So now we shall use this formula. Let's try using it to find a Fibonacci number we would ordinarily not find by using the recursive definition of the Fibonacci numbers—that is, writing out the Fibonacci sequence until we get to the 128th number. Applying the Binet formula for $n = 128$ gives us:

$$F_{128} = \frac{1}{\sqrt{5}}\left[\phi^{128} - \left(-\frac{1}{\phi} \right)^{128} \right] = \frac{1}{\sqrt{5}}\left[\left(\frac{1+\sqrt{5}}{2} \right)^{128} - \left(\frac{1-\sqrt{5}}{2} \right)^{128} \right]$$

$$= 251,728,825,683,549,488,150,424,261$$

(You might want to check this with the listing in appendix A.)

Thus we have the *Binet formula*:

$$F_n = \frac{1}{\sqrt{5}}\left[\phi^n - \left(-\frac{1}{\phi} \right)^n \right] = \frac{1}{\sqrt{5}}\left[\left(\frac{1+\sqrt{5}}{2} \right)^n - \left(\frac{1-\sqrt{5}}{2} \right)^n \right]$$

which will give us any Fibonacci number for any natural number n.

Let's stop and marvel at this wonderful result. For any natural number n the irrational numbers in the form of $\sqrt{5}$ seem to disappear in the calculation and a Fibonacci number appears. In other words, the Binet formula delivers the possibility of obtaining any Fibonacci number with the aid of the golden ratio ϕ.

At this point you may be wondering about the practicality of the Binet formula. Well, you are right if you surmised that we would likely obtain these larger Fibonacci numbers with the aid of a computer. However, there is an intrinsic value to knowing that the Binet formula exists and that it is possible to get the Fibonacci numbers in two ways: explicitly and recursively—just as we saw earlier was the case with the square numbers.

The Binet Formula for Lucas Numbers

It would be only fitting that there be an analogous formula to find a Lucas number without listing the Lucas sequence up to the number sought. Using for Lucas numbers a similar argument as the one we used for the Fibonacci numbers, we can get the Binet formula from figure 9-4. There we see that when n is odd, the Lucas number appears in the difference expression $\phi^n - \dfrac{1}{\phi^n}$, and when n is even, the Lucas number appears in the sum expression $\phi^n + \dfrac{1}{\phi^n}$. We can state this mathematically (as we did before) with the following:

$$L_n = \phi^n + (-1)^n \frac{1}{\phi^n}$$

That means (see the Binet formula):

$$L_n = \phi^n + (-1)^n \cdot \left(\frac{1}{\phi}\right)^n = \phi^n + \left(-\frac{1}{\phi}\right)^n = \left(\frac{1+\sqrt{5}}{2}\right)^n + \left(\frac{1-\sqrt{5}}{2}\right)^n$$

(for all $n \in \mathbf{N}$).[4]

To check this "inductively," we can test it for a few values of n. Let us try one odd and one even number.

---

4. This means for all natural numbers. Yet, in this case, we will also include 0.

For $n = 3$, we get:

$$L_3 = \phi^3 + (-1)^3 \frac{1}{\phi^3}$$

$$= \phi^3 + \left(-\frac{1}{\phi}\right)^3$$

$$= \phi^3 - \frac{1}{\phi^3} = 4$$

(We can use the chart in figure 9-4, rather than doing the computation again.)

For $n = 6$, we get:

$$L_6 = \phi^6 + (-1)^6 \frac{1}{\phi^6}$$

$$= \phi^6 + \frac{1}{\phi^6} = 18$$

Just to get a better feeling about how to evaluate these beyond the chart, we shall evaluate the Binet formula for the eleventh Lucas number, L_{11}.

$$L_{11} = \phi^{11} + (-1)^{11} \cdot \left(\frac{1}{\phi}\right)^{11} = \phi^{11} - \frac{1}{\phi^{11}}$$

$$= \left(\frac{\sqrt{5}+1}{2}\right)^{11} - \left(\frac{\sqrt{5}-1}{2}\right)^{11} = \frac{89\sqrt{5}+199}{2} - \frac{89\sqrt{5}-199}{2}$$

$$= 199$$

which *is* the eleventh Lucas number. So you can see how we can get any Lucas number without generating the sequence to that point—which would be potentially a very tedious task! Thus we have a Binet formula for generating Lucas numbers as well as the Fibonacci numbers.

Finding Individual Fibonacci Numbers

Suppose you have just found the twenty-fifth Fibonacci number and you would like the twenty-sixth Fibonacci number—without knowing the twenty-fourth Fibonacci number. We could use the Binet formula, but there is another way that might be simpler. Since we know that the ratio of two consecutive Fibonacci numbers is approximately the golden ratio, $\phi \approx 1.618$. That is,

$$\frac{F_{26}}{F_{25}} \approx 1.61803399$$

Therefore, we get $F_{26} \approx 1.61803399 \cdot F_{25}$, and then we round off the answer. So in order for us to find the twenty-sixth Fibonacci number, we just multiply the twenty-fifth Fibonacci number by 1.61803399 and then round off the answer: $75,025 \cdot 1.61803399 \approx 121,393.0001$, which when rounded off is 121,393, which is the twenty-sixth Fibonacci number. You can easily see—by trying—that this method of finding a Fibonacci number, given only the one before it, can be used for the Fibonacci numbers after the first one—that is, for finding F_n, where $n > 1$.

Another Way to Find a Specific Fibonacci Number (Using a Calculator or a Computer)

Finding a specific Fibonacci number can be done in several ways, as we have shown so far in this chapter. However, the use of a computer or a calculator presents us with yet another option. If we are asked to find the thousandth Fibonacci number, F_{1000}, then by using the definition—even with the use of technology—it would be a laborious task. We would have to use the definition: $F_0 = 0$; $F_1 = 1$; and $F_n = F_{n-1} + F_{n-2}$, where $n > 1$, and then find all 999 numbers that precede F_{1000}.

Consider the following two relationships:[5]

$$F_{2n-1} = F_{n-1}^2 + F_n^2 \text{ and } F_{2n} = F_n\left(2F_{n-1} + F_n\right)$$

With these two relationships, we need only F_n and F_{n-1} to compute both F_{2n} and F_{2n-1}. So here is how we would go about finding F_{1000}, using these relationships (with technological assistance naturally):

To find F_{1000}, we must first find F_{500} and F_{499}.

To find F_{500} and F_{499}, we must first find F_{250} and F_{249}.

To find F_{250} and F_{249}, we must first find F_{124} and F_{125}.

To find F_{125} and F_{124}, we must first find F_{61}, F_{62}, and F_{63}.

To find F_{61}, F_{62}, and F_{63}, we must first find F_{30}, F_{31}, and F_{32}.

To find F_{30}, F_{31}, and F_{32} we must first find F_{14}, F_{15}, and F_{16}.

To find F_{14}, F_{15}, and F_{16} we must first find F_6, F_7, and F_8.

To find F_6, F_7, and F_8 we must first find F_2, F_3, and F_4.

To find F_2, F_3, and F_4 we must first find F_0, F_1, and F_2.

Remember, $F_1 = F_2 = 1$, and $F_0 = 0$.

Although we had to compute 22 Fibonacci numbers, it is still a lot less work than having to compute 999 Fibonacci numbers to get the thousandth Fibonacci number.

Testing for Fibonacci Numbers

Now let's look at the situation in reverse. Suppose we were given a number and we wish to determine if this number is a Fibonacci number. There is (curiously enough) a method for testing if a number is a member of the Fibonacci sequence. The test goes this way:

---

5. The first of these relationships was considered in chapter 1, item 9 on page 43. The second relationship is new to us but also holds true. For a proof of this, see appendix B.

A number n is a Fibonacci number if (and only if)
$5n^2 + 4$ or $5n^2 - 4$ is a perfect square.[6]

We do this without a proof, but we can test this with some of our familiar numbers, say, at first, $n = 5$. We can then calculate that $5 \cdot 5^2 - 4 = 125 - 4 = 121$, which is a perfect square, namely, 11^2. Therefore, 5 is a Fibonacci number, since it passes this (square root) test. Indeed, we know $5 = F_5$. Now we choose $n = 8$. We can then calculate that $5 \cdot 8^2 + 4 = 320 + 4 = 324$, which is a perfect square, namely, 18^2. Therefore, 8 is a Fibonacci number since it passes this "square root" test. Indeed, we know that $8 = F_6$. We should also note that since 4 is not a Fibonacci number, it will fail this square root test. That is, $5n^2 \pm 4$ is not a perfect square, when $n = 4$: $5 \cdot 4^2 + 4 = 84$, and $5 \cdot 4^2 - 4 = 76$. You might want to try a few more examples to convince yourself that this powerful test really works.

We have now seen that the Fibonacci numbers—which are so closely tied to the sequence from which they emanate—can also be treated as individual entities. We can now identify specific Fibonacci numbers and can test to see if a given number is, in fact, a member of the Fibonacci number sequence.

---

6. This means the following: if $5n^2 + 4$ or $5n^2 - 4$ is a perfect square, then n is a Fibonacci number. And conversely, if n is a Fibonacci number, then $5n^2 + 4$ or $5n^2 - 4$ is a perfect square.

Chapter 10

The Fibonacci Numbers and Fractals

Unless you are young enough to have experienced recent reforms in school mathematics curricula, your idea of geometry is probably the study of ideal shapes such as lines, circles, squares, and rectangles. More precisely, you have probably studied the objects and mathematical relationships compiled by the Greek mathematician Euclid (ca. 365–300 BCE), from more than two thousand years ago. The so-called Euclidean geometry studies points, lines, planes, and objects such as circles and polygons—objects that for many of us precisely denote the term *geometrical figures*. We usually think of a geometrical figure when we recognize a circle on the surface of a wheel, or a rectangle on a tabletop. After all, it is not so difficult to find approximations of these figures in objects made by people. However, how often do we find smooth planes, lines, or curves in nature?

Natural objects are often irregular. Their surfaces and lines are usually rough and fragmented. For example, compare the smooth feel of a finished wooden table to the roughness of a tree bark. Or compare the prickly crown of a real-life conifer, such as a pine tree, to the solid shape that names it, the cone. Or, moreover, think of the appearance of ragged coastlines, with all their projections

---

This chapter was written by Dr. Ana Lúcia B. Dias, professor of mathematics at Central Michigan University.

and indentations, in comparison to that of uniform curves such as the circle, in which all points in its circumference lie regularly at a same distance from the center. We can see that although Euclidean geometry is useful in human creations, it is definitely not the best set of tools to describe and understand natural objects or phenomena.

In the late nineteenth and early twentieth centuries mathematicians such as Gaston Julia (1893–1978), Pierre Fatou (1878–1929), and Georg Cantor (1845–1918) experimented with curves that have revealed themselves to be much more suited to represent and model natural phenomena. Although their contemporaries were not very enthusiastic about the new objects, referring to them as monstrous and pathological, mathematicians and nonmathematicians alike for decades now have been fascinated by the beauty of images of fractals made possible by computer graphics (figure 10-1).

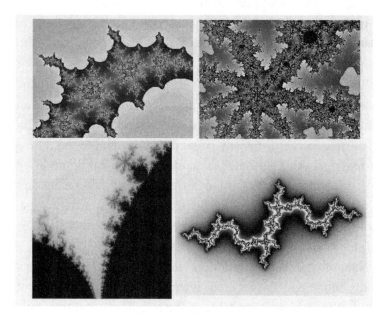

Figure 10-1
Some images of fractals.

With the aid of the computer, mathematician Benoît Mandelbrot (b. 1924) showed in 1980 that pictures of the mathematical objects created by Gaston Julia were quite beautiful, not monstrous. And more important, he showed that, rather than pathological, the ragged outlines and the repeating patterns of those figures were often found in nature (figure 10-2). He used the Latin word *fractus*, meaning broken or fractured, to coin a word to denote the new mathematical objects: *fractals*.

Figure 10-2

In figure 10-2 the images to the left are pictures of real-life scenes. Those to the right are fractal models.

Interestingly, the Fibonacci numbers also arise in the analysis of some fractals. In this chapter we are going to describe two instances in which Fibonacci numbers appear in two different fractals, even though the construction of the Fibonacci sequence was by no means embedded in the procedures for the creation of such fractals—which makes the finding of such numbers all the more fascinating!

The characteristic feature of fractals is self-similarity: geometric patterns seen in the big picture of a fractal are repeated in its parts in smaller and smaller scales.

Making fractals involves the repeated application of a geometric rule or transformation to an original figure or set of points, which we will refer to as the *seed* of the fractal.

Once we determine what the generative procedure and the seed of a fractal will consist of, we can begin constructing the fractal by repeatedly applying the generative procedure—first to the seed, then once again to the resulting output, and so on. Therein lies another definitive aspect of fractal construction: fractals are made up of consecutive phases called *iterations*. An iteration is the act of applying one algorithm or procedure one time through in a repetitive process.

When constructing a fractal, the iterations of the generative procedure are done recursively; that is, the input of each iteration is the output of the previous one—with the exception of the first iteration, which is applied to a seed. In some cases this means that each subsequent iteration will be more cumbersome than the previous one! In those cases, programmable technology is definitely a great help.

A fractal ideally entails the iteration of a procedure an infinite number of times, although in practice we can iterate a procedure only a finite number of times. We can use computers or calculators to help us perform as many iterations as we want, which would give us different stages in the construction of a fractal. Or we can use mathematics to deduce what would be the result of performing that infinite process.

Before we move on to fractals that yield the Fibonacci numbers, we will illustrate the generative process described above and the appropriate terminology through a classical example, the Koch snowflake (1904)[1](figure 10-3).

For the construction of this fractal, the seed will be an equilateral triangle. Because the generation of the fractal happens by suc-

---

1. Named after the Swedish mathematician Helge von Koch (1870–1924).

cessive iterations, we will call the result of each iteration a *stage* in the fractal construction.

The generative procedure will consist of erasing the middle third of every line segment in a stage and replacing it with two line segments of the same length (one-third of the original segment) at an angle of 60°; this will form cusps (which look like partial equilateral triangles), where before there were segments. We can see this procedure illustrated in figure 10-3.

Figure 10-3
Generative procedure for the Koch snowflake.

Each iteration will consist of applying the fractal-construction procedure to each line segment in a stage of the fractal, which will create the next stage. Figure 10-4 shows the first two iterations in the construction of the Koch snowflake. But now we do this on three sides of an equilateral triangle.

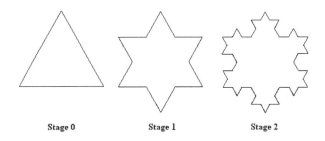

Stage 0   Stage 1   Stage 2

Figure 10-4
Construction of Koch snowflake.

You may want to sketch stage 3 of the Koch snowflake on a separate piece of paper, or use a computer drawing program. Remember this iteration will require applying the generative procedure to every segment in stage 2. It involves quite a bit more work than the previous iteration. While the first iteration consisted of applying the generative procedure to three line segments, in the second iteration that number increased to twelve. In the third iteration you will need to work on forty-eight line segments. That increase in complexity is displayed in figure 10-5. With each iteration, each line segment at a stage will be replaced by four line segments, forming a cusp, in the next stage. So if we know how many segments there are at a stage, we can find out the number of segments at the next stage by multiplying that number by four. This relation can be written algebraically and recursively as: $S_n = 4 \cdot S_{n-1}$ (which just happens to equal $3 \cdot 4^n$), where S_n is the number of segments at stage n, and S_{n-1} is the number of segments at the previous stage.

Stage (n)	Number of Line Segments (S_n)
0	3
1	$12 = 4 \cdot 3$
2	$48 = 4 \cdot 12$
3	$192 = 4 \cdot 48$
n	$S_n = 4 \cdot S_{n-1}$

Figure 10-5

The replacement of each segment by a visual spike, and the accelerating increase in the number of segments in this fractal, is what gives it the main features of a fractal: the jagged appearance and the property of self-similarity. If you zoom in on any spike, you will find smaller and smaller copies of it. Magnifying fractals reveals in them small-scale details similar to the large-scale characteristics.

Another popular fractal is the Sierpinski[2] gasket (1915) (figure 10-6). Its seed is also an equilateral triangle. Each iteration consists of splitting a triangle into four smaller equilateral triangles by using the midpoints of the three sides of the original triangle as the new vertices, then deleting the middle triangle from further action,

---

2. Named after Waclaw Sierpinski (1882–1969), a Polish mathematician.

that is, a quarter of the area is removed. The construction of the fractal continues by iterating this procedure over and over: from every triangle formed, a triangle is removed from its interior. This results in not only a rough, fragmented surface, but also in self-similarity—the two main features of fractals.

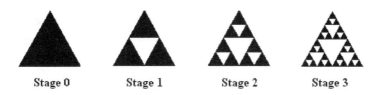

Stage 0   Stage 1   Stage 2   Stage 3

Figure 10-6
Construction of the Sierpinski gasket.

We are now ready to encounter our protagonist, the Fibonacci numbers. In fractals, as well as elsewhere, these numbers make intriguing incidental appearances.

One of such unexpected sightings of the Fibonacci sequence occurred in the examination of a recently created fractal, the Grossman Truss.[3] Let us see how this fractal is constructed: we start with an isosceles right triangle as the seed. An isosceles right triangle has two sides of equal length forming a right angle (figure 10-7).

Figure 10-7
Stage 0 of the construction of the Grossman truss.

---

3. The creation of this fractal and a mathematical analysis of its properties were reported by George W. Grossman, in "Construction of Fractals by Orthogonal Projection Using Fibonacci sequence," *Fibonacci Quarterly* 35, no. 3 (August 1997): 206–24.

We will now drop a perpendicular from its right angle until it intersects the opposite side. We then bounce that perpendicular clockwise, until it meets the opposite side, again orthogonally or perpendicularly. Finally, we will delete the triangle created at the interior of the figure. This completes the first iteration of the generative procedure. The result is a new figure, in which the original isosceles right triangle appears split into two other triangles, with a triangular gap between them. If you examine this figure carefully, you will see that all triangles formed are also isosceles right triangles—that is, they are similar to the original seed, but they appear in different positions (figure 10-8).

Figure 10-8
Stage 1 of the construction of the Grossman truss.

It's time to move on to the second iteration. In this fractal, each iteration consists of repeating the procedure described above in the largest triangles at each stage. Take a look at the triangles still existing on Stage 1 (remember the middle triangle has been deleted and is no longer part of the figure—a blank remains in its place). You can see that there are two triangles of different size. Therefore, the next iteration will consist of applying the generative procedure to the largest and only to the largest triangle. The result is shown in figure 10-9.

Figure 10-9
Stage 2 of the construction of the Grossman truss.

Now that our fractal has another gap in it, what can we say about the remaining triangles? Are they all equally sized? Careful inspection of the shaded triangles in figure 10-9 reveals that there are two sizes of triangles at this stage: two large ones and one small one. So for the next iteration, the generative procedure will need to be performed on the two large triangles.

The construction of this fractal proceeds in this way, always applying the generative procedure to the largest triangles at a stage. The result is a nice-looking fractal, in which the basic shape shown in figure 10-9 can be repeatedly seen in smaller scales and at different positions throughout the fractal. Figure 10-10 shows the fractal after eight iterations.

Figure 10-10
Stage 8 of the construction of the Grossman truss.

In figure 10-10, as well, we can see that the remaining triangles come in two sizes. Can you tell how many triangles there are of the larger and how many of the smaller sizes? Maybe it will help to look at the fractal after fewer iterations. Figure 10-11 shows the Grossman truss after four iterations. Again, there are shaded triangles of two sizes. They have been filled with different shades of gray to help us count.

Figure 10-11
The Grossman truss after 4 iterations.

If we look at the shaded triangles in figure 10-11, we see five larger triangles and three smaller triangles. These numbers happen to be the fifth and fourth terms in the Fibonacci sequence.

What happens after one more iteration? The number of larger triangles is now eight, and there are five smaller triangles (figure 10-12).

Figure 10-12
Stage 5 of the construction of the Grossman truss, showing the eight larger triangles in the dark shade of gray and the five smaller triangles in the lighter shade.

With trained eyes, you should be able to find that in figure 10-10 there are twenty-one smaller triangles and thirty-four larger ones. This pattern continues indefinitely—the number of triangles of a like size in the Grossman truss is always a Fibonacci number!

There are more places where the Fibonacci numbers appear. We can also shift our gaze to the triangles that have been deleted from the fractal—the empty triangular gaps. How many sizes are there at a determined stage, and how many of each size?

Figure 10-13 shows three copies of Stage 3 of the Grossman truss. How many different-sized triangular gaps are there in Stage 3? There is one gap of the largest size; one gap is the second-largest; and there are two of the smaller gaps. Do you see a pattern being formed?

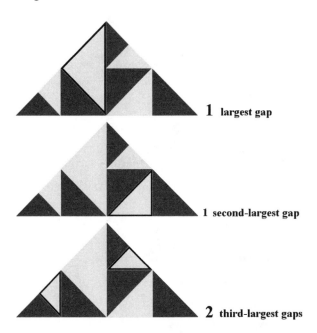

Figure 10-13
Counting gaps in Stage 3 of the Grossman truss.

Can you predict how many different-sized gaps there will be after one more iteration? Remember, all the gaps at a stage will still be there in subsequent iterations, since the gaps are removed from further action. But the procedures applied to the areas of the fractal that are still filled will create more gaps, of smaller and smaller sizes. How many new gaps will be created with each iteration?

The pattern suggested by figure 10-13 becomes even more striking if we look at further iterations. Figure 10-14 shows copies of Stage 5 of the fractal, and a count of the number of gaps of each different size.

Figure 10-14
Counting gaps in Stage 5 of the Grossman truss.

Again, it is the Fibonacci sequence! And whereas counting triangles of the Grossman truss gives us two Fibonacci numbers at every stage, counting gaps can give us the entire sequence in a single stage. More precisely, counting gaps with the Grossman truss after *n* iterations will give us the *n* first terms of the Fibonacci sequence!

What is most amazing about this fractal is that the Fibonacci sequence emerges in it even though it was not directly involved in the procedure that generated it.

The way in which the Grossman truss was created did not hint at a possible participation of the Fibonacci numbers in the resulting fractal. This makes all the more mysterious and amazing the ways in which this sequence appears both in nature and in human-created objects—such as fractals. It makes us wonder about the true nature of mathematics: even though mathematical objects are human creations, the relationships within them seem to follow by necessity. They emerge even though we did not put them there.

Robert Devaney[4] has shown another amazing way of finding Fibonacci numbers in a fractal. The Fibonacci sequence has made its way into one of the most well-known fractals: the Mandelbrot set (figure 10-15).

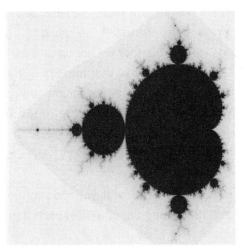

Figure 10-15

4. A mathematical exposition of these concepts is reported in Robert L. Devaney, "The Fractal Geometry of the Mandelbrot Set," in *Fractals, Graphics, and Mathematics Education*, ed. Michael Frame and Benoît Mandelbrot (Washington, DC: Mathematical Association of America, 2002), pp. 61–68

But first, let's see what the Mandelbrot set is. Its image is so popular it could earn the title of the "emblem of fractal geometry." Its strange beauty mesmerizes laymen and experts alike. But what does that image represent? As with the other fractals we have examined, some elements are involved in its construction: a seed, a rule or transformation, and an infinite number of iterations. But unlike our previous examples—which were mainly geometrical—the Mandelbrot set is a set of numbers. The image you see in figure 10-15 is just a plot, in the complex plane,[5] of the numbers that belong to the set.

How do we know whether a number is or is not in the Mandelbrot set? We have to test each number to find out. This infinitely large task can only be done with the aid of a computer, and only a finite number of times, although a very large number of times. In fact, it was only under the right conditions, in which Benoît Mandelbrot's vision and intellect was combined with the environment of IBM's Watson Research Center, that a revival of the work on this set that had been initiated by Julia in the 1920s was made possible.

So the construction of the image of the Mandelbrot set requires one more element besides the seed, the rule, and the iterations that the fractals, previously discussed, also had: it involves a test of numbers. Let us say the number we are testing is c.

The seed for this fractal is the number 0; not a triangle or a segment, but a number, because this fractal is numerical in nature. The rule or transformation is: **square the input and add c**, which can be expressed algebraically as $x^2 + c$.

Suppose we want to test the number $c = 1$. Our transformation becomes: $x^2 + 1$.

Let us see the result of a few iterations, starting with the seed 0 as the input, and then using the output of each iteration as the input for the next:

$$0^2 + 1 = 1$$
$$1^2 + 1 = 2$$
$$2^2 + 1 = 5$$
$$5^2 + 1 = 26$$
$$26^2 + 1 = 677$$
$$677^2 + 1 = 458,330$$

We can see that the more iterations, the greater the result will be. The terms of the sequence of numbers will increase without bound. We say that "it goes to infinity."

Let us test for another number, $c = 0$. With this value for c, our rule becomes: $x^2 + 0$.

Starting with the same seed 0, a few iterations will show that the sequence will be fixed at 0:

First iteration: $0^2 + 0 = 0$
Second iteration: $0^2 + 0 = 0$

For each value of c, the "test" (repeatedly iterating the rule) will tell us whether the result will go to infinity, or if it will not. Values of c that will result in an escape to infinity are *not* in the set; all the others *are* in the set. The image of the Mandelbrot set is actually a record of the fate of each number, c, under this test.[6] The key to understanding the image is to unveil the code used. The most frequently used code for plotting the results of these tests is to use the color black to represent those points in the plane that *are* in the Mandelbrot set and to color the others according to their "escape speed," that is, using different colors

6. In fact, figure 10-15 is only an approximation of the Mandelbrot set. In actuality, we cannot know for sure whether a number c lies in the Mandelbrot set, because to determine that with absolute certainty we would need to iterate the "test" an infinite number of times. But even with computers, we can obviously iterate anything only a finite number of times. But it so happens that the sequence formed by iterating the rule to a certain value of c may behave differently only after a very large number of iterations. So we can make our approximation better by iterating a great number of times. Still, this will not lead to absolute accuracy.

to represent the number of iterations that value takes to reach a certain distance from the origin. Another traditional way of plotting the Mandelbrot set is just to use black for points that are in the set and white for those that are not.

We will now look at the image of the Mandelbrot set with a categorical eye. At the core of the image, we can see a heart-shaped figure, the *main cardioid*.[7] We can also note many round decorations, or *bulbs* (figure 10-16). We call any bulb that is directly attached to the main cardioid a *primary bulb*. The primary bulbs have in turn many smaller decorations attached to them. Among them, we can identify what appear to be *antennas* (figure 10-17).

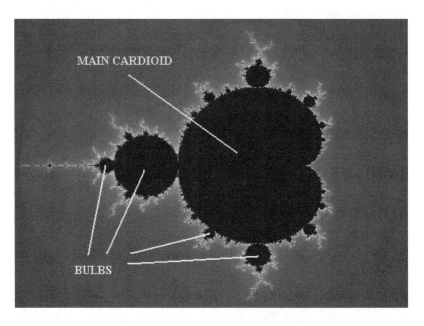

Figure 10-16
The main cardioid and bulbs in
the Mandelbrot set.

---

7. A cardioid is a heart-shaped curve generated by a fixed point on a circle as it rolls around another circle of equal radius.

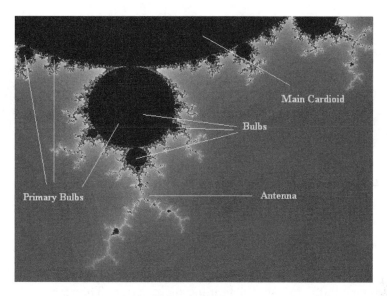

Figure 10-17
Detail of decorations in the Mandelbrot set.

We will call the longest of these antennas the *main antenna*. Finally, the main antennas show several "spokes" (figure 10-18). Note that the number of spokes in a main antenna varies from decoration to decoration. We will call this number the *period* of that bulb or decoration. To determine the period of that bulb, just count the number of spokes on an antenna. Don't forget to count the spoke emanating from the primary decoration to the main junction point. Figure 10-19 displays various primary bulbs and their periods.

Figure 10-18
Main antennas and their "spokes."

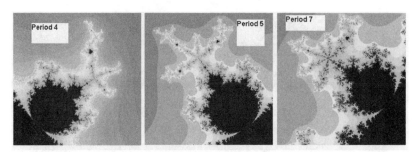

Figure 10-19
**Determining the period of bulbs by counting the "spokes" in
their main antenna.**

How can the Fibonacci sequence be seen in the Mandelbrot set? We will consider the period of the main cardioid to be 1. Then, by counting the spokes in the main antenna of the largest primary bulbs, we will determine their period. The results of this counting, that is, the period of the main cardioid and of some of the largest primary bulbs, are registered in figure 10-20.

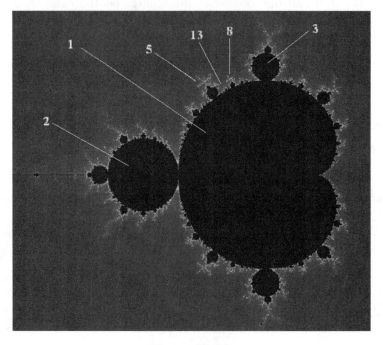

Figure 10-20
The Fibonacci numbers in the Mandelbrot set.

You may be surprised to see from inspection of figure 10-18 that the largest bulb between the bulb of period 1 and the bulb of period 2 is a bulb with period 3. The largest bulb between a period-2 bulb and a period-3 bulb is a period-5 bulb. And the largest bulb between a period-5 bulb and a period-3 bulb is a period-8 bulb. Voila! Once again the Fibonacci numbers appear. Why is that so? There are no obvious explanations. The Fibonacci numbers are not directly related to the way in which the periods of primary bulbs are calculated. Once again, the Fibonacci sequence makes a mysterious and remarkable appearance.

So where else might the Fibonacci sequence emerge? Apparently, if we develop the habit of counting—counting the most diverse sets of objects we can think of: seeds, spikes, triangles—we are bound to find the Fibonacci sequence in many of them.

Should you develop the habit of counting various collections of things, you will probably make your own amazing sightings of the Fibonacci sequence. It is mostly everywhere—all we have to do is to look carefully.

Epilogue

Now that you have experienced the wonders of what might be considered the most famous sequence of numbers in all of mathematics, you will undoubtedly be keen to search for other manifestations of them. You should be aware that there are constantly new discoveries made where the Fibonacci numbers amazingly appear. Even if some might argue that a few of our examples might be a bit forced or contrived, no one can deny how ubiquitous the presence of this sequence of numbers is.

We found the Fibonacci numbers in nature and art. They grew to prominence in the seminal work by Leonardo of Pisa—Fibonacci, who brought the ten numerals to the attention of Western culture. The Fibonacci numbers have been found to hold structures together in the most beautiful way as an integral part (or generator) of the golden ratio, and they have tied together more diverse branches of mathematics than perhaps any other entity in mathematics. They even manifest themselves in the way we manipulate investments on the stock market.

Not only do the Fibonacci numbers demonstrate what most mathematicians vehemently argue is the beauty in mathematics, but they also seem to be either an integral part of what we in our society consider beautiful or they help determine that beauty. This can be seen in the many art works, both old and new, that exhibit the golden ratio—itself determinable from the Fibonacci numbers. Though for some artists and composers the golden ratio plays less of a role than for others—and, of course, music and art are more than a mathematical formula. Its significance in the arts is undeniable. Whether it is the beauty of these numbers or the challenge they continuously lay before us to discover new manifestations of their appearance, the Fibonacci numbers never cease to amaze us. Mathematicians, through the Fibonacci Association, established in 1963, continue to find applications of these numbers both inside and outside the field of mathematics and publish them four times a year.

The Fibonacci numbers are very easy to remember, but if you cannot commit, say, the first twelve to memory (remember F_{12} just happens to be equal to $12^2 = 144$), we have the additive process to rely on, and for very large Fibonacci numbers, the convenient Binet formula will bring us there immediately. As you look for the Fibonacci numbers in any field, remember, care must be taken not to overstate their application. Not all beauty is consistent with the Fibonacci numbers—as we have found in music—and not all art relies on the golden ratio for its beauty. Yet there is a role for the Fibonacci numbers in both of these fields as long as we observe it objectively and not force their appearance. There are plenty of natural illustrations, so we do not have to artificially claim the numbers' presence artificially.

We hope, by reading this book, you will not only have been motivated to search for new appearances of the Fibonacci numbers, but will also have developed a new, or renewed, love for mathematics and a greater appreciation of the beauty of this most important science—referred to by the famous German mathematician Carl Friedrich Gauss (1777–1855) as "the queen of science."

Afterword

By
Herbert A. Hauptman

The Fibonacci numbers have fascinated me for decades. Yet when I was asked to provide my thoughts about them, I decided to see if I could discover patterns independent of what others may have found earlier. One never knows if his discoveries are new, for when one asks mathematicians if some of these discoveries are new, one often gets the answer: "Well, I haven't seen this before, but that doesn't mean that it hasn't been published somewhere." Since 1963 the Fibonacci Association (mentioned elsewhere in this book) has published a quarterly journal on the seemingly endless relationships involving the Fibonacci numbers. I offer here some of my "discoveries" in the hope that they will provide further insights and motivate the reader to seek others.

1. Some Plausible Speculations

We begin with what must surely be one of the most fundamental concepts of all: the idea of unity, or the number one. We continue in this vein with what is, just as certainly, the most primitive arithmetical operation: the operation of addition. We obtain in this way the subject matter of this book, a construct of great beauty, the wonderful sequence of numbers known as the Fibonacci numbers, 1, 1, 2, 3, 5, 8, 13, ... As the reader already knows, each number in this sequence, after the first two, is obtained by addition of the preceding two. Can one imagine a simpler construction? Yet this sequence of numbers has the most remarkable properties, many of which are described in this book. In this afterword, adopting a somewhat different point of view, we give a brief overview of the divisibility properties of the Fibonacci numbers and hint at the exciting further developments, which the reader is invited to explore.[1]

1.1. Notation

Denote by F_n the nth Fibonacci number, so that $F_1 = 1$, $F_2 = 1$, $F_3 = 2$, $F_4 = 3$, $F_5 = 5$, ...

1.2. The Even Fibonacci Numbers

We ask a simple question. Which of the Fibonacci numbers are divisible by 2? Consulting the table of Fibonacci numbers, we observe that F_3, F_6, F_9, F_{12}, F_{15}, ... are all divisible by 2 and, in fact, these are the only even Fibonacci numbers. We are tempted therefore to conjecture that the even Fibonacci numbers consist of the set F_3, F_6, F_9, F_{12}, F_{15}, ... We note in addition that the subscripts 3, 6, 9, 12, ... of the even Fibonacci numbers themselves form an arithmetic progression with initial term and common difference equal to 3; in other words, it consists of all the numbers divisible by 3.

Encouraged by this discovery, we ask next which Fibonacci numbers are divisible by 3.

1.3. The Fibonacci Numbers That Are Divisible by 3

Again consulting the table, we observe that those Fibonacci numbers that are divisible by 3 are $F_4 = 3$, $F_8 = 21$, $F_{12} = 144$, $F_{16} = 987$, $F_{20} = 6,765$, ... The subscripts 4, 8, 12, 16, 20, ... again form an arithmetic progression, this time with the initial term and common difference equal to 4, and therefore consist of all numbers that are

---

1. This was briefly mentioned in chapter 1 (pages 47–48, item 12)

divisible by 4. We conjecture therefore that the collection of the subscripts of all the Fibonacci numbers that are divisible by 3 themselves consist of all numbers divisible by 4.

1.4. The Fibonacci Numbers Divisible by 4

The pattern is now clear. We now find the Fibonacci numbers divisible by 4 to be $F_6 = 8$, $F_{12} = 144$, $F_{18} = 2,584$, $F_{24} = 46,368$, . . . , the subscripts of which form an arithmetic progression with initial term and common difference equal to 6. Hence the subscripts of all Fibonacci numbers that are divisible by 4 are themselves divisible by 6 and in fact consist of all the multiples of 6.

Proceed as in 1.2, 1.3, and 1.4 to deduce the following: the Fibonacci numbers F_n that are divisible by 5 (or 6, or 7, respectively) have subscripts, n, that are divisible by 5 (or 12, or 8, respectively) and, in fact, consist of all the multiples of 5 (or 12, or 8, respectively).

2. The Minor Modulus $m(n)$

In this section we formulate in greater generality what we have learned in the previous section. Let n be any positive integer. In view of our previous work, we make the fundamental assumption that there exist infinitely many positive integers x such that the Fibonacci number F_x is divisible by n. The smallest such number x clearly depends on n. We call it the minor modulus of n and denote it by $m(n)$. Thus $m(n)$ is the smallest positive integer x such that F_x is divisible by n. Referring to our previous work, we deduce, for example, that $m(2) = 3$, $m(3) = 4$, $m(4) = 6$, $m(5) = 5$, $m(6) = 12$, and $m(7) = 8$. A more extensive table of values of the minor modulus $m(n)$ is readily derived as explained in the previous section, and is shown in table 1. It is assumed that the reader has extended this table to $n = 200$ at least so he can confirm to his satisfaction the several conjectures made here.

Table of $m(n)$, $1 \leq n \leq 100$

n	$m(n)$	n	$m(n)$	n	$m(n)$	n	$m(n)$	n	$m(n)$
1	1	21	8	41	20	61	15	81	108
2	3	22	30	42	24	62	30	82	60
3	4	23	24	43	44	63	24	83	84
4	6	24	12	44	30	64	48	84	24
5	5	25	25	45	60	65	35	85	45
6	12	26	21	46	24	66	60	86	132
7	8	27	36	47	16	67	68	87	28
8	6	28	24	48	12	68	18	88	30
9	12	29	14	49	56	69	24	89	11
10	15	30	60	50	75	70	120	90	60
11	10	31	30	51	36	71	70	91	56
12	12	32	24	52	42	72	12	92	24
13	7	33	20	53	27	73	37	93	60
14	24	34	9	54	36	74	57	94	48
15	20	35	40	55	10	75	100	95	90
16	12	36	12	56	24	76	18	96	24
17	9	37	19	57	36	77	40	97	49
18	12	38	18	58	42	78	84	98	168
19	18	39	28	59	58	79	78	99	60
20	30	40	30	60	60	80	60	100	150

Table 1

We summarize our major conjecture as follows: Let n be any positive integer; denote by $m = m(n)$ the minor modulus of n. Then the Fibonacci number F_x is divisible by n if and only if x is divisible by m.

The minor modulus $m(n)$ itself has a number of interesting properties that inspection of table 1 makes plausible and which we summarize shortly. First however we digress to give a brief account of the prime numbers that play a special role in this development.

3. The Primes

Definition: Every number (> 1) is divisible by itself and unity. If these are its only divisors, it is said to be prime. Thus, 17, which is divisible by no numbers other than 1 and 17, is prime. Similarly, 2, 3, 5, 7, 11, and 13 are also primes. A famous theorem attributed to Euclid states that there are infinitely many primes. Numbers like $6 = 2 \cdot 3$ or $8 = 2^3$ are, on the other hand, not primes since 6 is divisible by 2 (in addition to 1, 3, and 6) and 8 is divisible by 2 (in addition to 1, 4, and 8). They are said to be composite. The reader may readily confirm that the first few composite numbers are 4, 6, 8, 9, 10, and 12. Clearly all primes, other than 2, are odd.

4. $m(p^k)$ Where p Is Prime and k Is a Positive Integer

The minor modulus of a prime power is of particular importance in the sequel. Reference to table 1 makes plausible the following conjectures:

4.1.
$$m(2^1) = m(2) = 3, k = 1$$
$$m(2^2) = m(4) = 6, k = 2$$
$$m(2^k) = 3 \cdot 2^{k-2}, \text{ if } k > 2$$

4.2.
$$m(5^k) = 5^k$$

4.3.
$$\text{If } p \text{ is an odd prime, } m(p^k) = p^{k-1} m(p)$$

For example, to confirm 4.1, observe that

when $k = 1$, $F_3 = 2$, which is divisible by 2, and $x = 3$ is the smallest value of x for which F_x is divisible by $2^1 = 2$, whence, by definition, $m(2) = 3$;

when $k = 2$, $F_6 = 8$, which is divisible by 2^2, and $x = 6$ is the smallest value of x for which F_x is divisible by $2^2 = 4$, whence $m(2^2) = 6$;

when $k = 3$, $F_6 = 8$, which is divisible by 2^3, and $x = 6$ is the smallest value of x for which F_x is divisible by $2^3 = 8$, whence $m(2^3) = 6$;

when $k = 4$, $F_{12} = 144$, which is divisible by 2^4, and $x = 12$ is the smallest value for x for which F_x is divisible by $2^4 = 16$, whence $m(2^4) = 12$;

when $k = 5$, $F_{24} = 46{,}368$, which is divisible by 2^5, and $x = 24$ is the smallest value of x for which F_x is divisible by $2^5 = 32$, whence $m(2^5) = 24$;

when $k = 6$, $F_{48} = 4{,}807{,}526{,}976$, which is divisible by 2^6, and $x = 48$ is the smallest value of x for which F_x is divisible by $2^6 = 64$, whence $m (2^6) = 48$;

and so on.

Next, to confirm 4.2, observe that when $k = 1$, $F_5 = 5$, which is divisible by 5, and $x = 5$ is the smallest value of x for which F_x is divisible by 5.

Whence, by definition, $m(5) = 5$, and when $k = 2$, $5^2 = 25$, $F_{25} = 5^2 \cdot 3{,}001$, which is divisible by 5^2, and $x = 25$ is the smallest value of x for which F_x is divisible by 5^2.

Whence, again by definition, $m(5^2) = 5^2$, and when $k = 3$, $5^3 = 125$, $F_{125} = 5^3 \cdot 3{,}001 \cdot 158{,}414{,}167{,}964{,}045{,}700{,}001$, which is divisible by 5^3, and $x = 125$ is the smallest value of x for which F_x is divisible by 5^3.

Whence, by definition, $m(5^3) = 5^3$, and when $k = 4$, $5^4 = 625$, $F_{625} = 5^4 \cdot P$, where P is a product of 5 primes, which is divisible by 5^4.

Whence, by definition again, $m(5^4) = 5^4$, and so on.

Finally, we shall confirm 4.3 for odd primes, $p = 3, 5, 7$ and $k = 1, 2, 3$, and leave to the reader the pleasure of confirming 4.3 for primes 11, 13, 17, 19 and $k = 1, 2, 3$.

Thus when $p = 3, k = 2$ we have from table 1 and the Fibonacci table:

$$m(3) = 4, \ m(3^2) = 12 = 3^1 \cdot m(3) = 3 \cdot 4$$

when $p = 3, k = 3$,

$$m(3) = 4, \ m(3^3) = 36 = 3^2 \cdot m(3) = 9 \cdot 4$$

when $p = 3, k = 4$,

$$m(3) = 4, \ m(3^4) = 108 = 3^3 \cdot m(3) = 27 \cdot 4$$

when $p = 5, k = 2$,

$$m(5) = 5, \ m(5^2) = 25 = 5^1 \cdot m(5) = 5 \cdot 5$$

when $p = 5, k = 3$,

$$m(5) = 5, \ m(5^3) = 125 = 5^2 \cdot m(5) = 25 \cdot 5$$

when $p = 5, k = 4$,

$$m(5) = 5, \ m(5^4) = 625 = 5^3 \cdot m(5) = 125 \cdot 5$$

when $p = 7, k = 2$,

$$m(7) = 8, \ m(7^2) = 56 = 7^1 \cdot m(7) = 7 \cdot 8$$

when $p = 7, k = 3$,
$$m(7) = 8, \ m(7^3) = 7^2 \ m(7) = 49 \cdot 8 = 392$$

and we readily confirm from the table that F_x is divisible by $7^3 = 343$ when $x = 392$ but for no smaller value of x.

5. Special Primes $p = 10n \pm 1$, $q = 10n \pm 3$ If n Is a Positive Integer

If, upon division by 10, the prime p leaves the remainder 1 or 9, it is said to be a prime $10n \pm 1$, because in this case there exists an integer n such that $p = 10n \pm 1$. For example, the first few primes $10n \pm 1$ are: $p = 11, 19, 29, 31, 41, 59, 61, 71$, and 79, since $11 = 10 \cdot 1 + 1$, $19 = 10 \cdot 2 - 1$, $29 = 10 \cdot 3 - 1$, $31 = 10 \cdot 3 + 1$, and so on.

Note that the last digit of a prime 10 ± 1 is either 1 or 9.

If, on the other hand, upon division by 10 the prime q leaves the remainder 3 or 7, it is said to be a prime $10n \pm 3$ because in this case there exists an integer n such that $q = 10n \pm 3$. For example, the first few primes $10n \pm 3$ are $q = 3, 7, 13, 17, 23, 37, 43, 47$, and 53.

Note that the last digit of a prime $10n \pm 3$ is either 3 or 7.

Every odd prime other than 5 is either $p = 10n \pm 1$ or $q = 10n \pm 3$. The special importance of primes $p = 10n \pm 1$ and primes $q = 10n \pm 3$ is due to the following, 5.1 and 5.2, which the reader is invited to confirm.

5.1. If p Is a Prime $10n \pm 1$ Then $m(p)$ Is a Divisor of $p - 1$.

Consult table 1 to confirm for the primes $p = 11, 19, 29, 31, 41, 59, 61, 71, 79$, and 89.

5.2. If q Is a Prime $10n \pm 3$ Then $m(q)$ Is a Divisor of $q + 1$.

Consult table 1 to confirm for the primes $q = 3, 7, 13, 17, 23, 37, 43, 47, 53, 67, 73, 83$, and 97.

6. A Simple Exercise

For each fixed positive integer $n = 2, 3, 4, \ldots$ consult table 1 to calculate the sequence $n, m(n), m(m(n)), m(m(m(n))), \ldots$.

What conclusion do you draw?

7. Greatest Common Divisor

We conclude this afterword with the remarkable formula described below in section 8. We shall need a short preliminary digression.

The greatest of all common divisors of two specified positive integers r and s is denoted by $g = (r, s)$.

It has the property that any common divisor of r and s is also a divisor of g.

Thus, if $r = 30$, and $s = 75$, then $g = (30, 75) = 15$. We see that 15 is the greatest of all common divisors of 30 and 75. Observe that the common divisor 5 of 30 and 75 is also a divisor of 15, but not the greatest one.

If r and s have no common divisor other than 1, we say that r and s are relatively prime and write $(r, s) = 1$.

8. A Remarkable Formula

We conclude with four remarkable properties of the Fibonacci numbers. In number 4 above, we described properties of $m(2^k)$ and $m(p^k)$, where p is an odd prime, which, together with table 1, facilitates the calculation of $m(n)$, when n is an arbitrary power of a prime. Here we give a formula which, again by means of table 1, facilitates the calculation of $m(n)$ for arbitrary integers n. This formula expresses $m(rs)$, where r and s are relatively prime, in terms of $m(r)$ and $m(s)$. The reader may readily confirm that if $(r, s) = 1$ then

$$m(rs) = \frac{m(r)m(s)}{(m(r),\ m(s))}$$

In other words, if r and s are relatively prime, that is, they have no common factor other than unity, then $m(rs)$ is equal to the product of $m(r)$ and $m(s)$ divided by the greatest common divisor of $m(r)$ and $m(s)$. Confirm this formula by consulting the extended table 1. Clearly this formula facilitates the calculation of $m(n)$ once the values of $m(2^k)$ and $m(p^k)$, where p is an odd prime, have been found.

Let me again summarize the ideas involving the minor modulus $m(n)$.

Let n be any positive integer. Then there exist infinitely many positive integers x such that the Fibonacci number F_x is divisible by n. The smallest such number x clearly depends on n. We call it the minor modulus of n and denote it by $m(n)$. Then $m(n)$ is the smallest positive integer x, such that F_x is divisible by n.

Let n be any positive integer. Then F_x is divisible by n if and only if x is divisible by $m(n)$. Alternatively, all solutions of the congruence

$$F_x \equiv 0 \ (mod \ n)$$

are given by

$$x \equiv 0 \ (mod \ m(n))$$

We also have for $m(p^k)$, where p is prime, the following:

1. $m(2) = 3$, $m(4) = 6$, if $k > 2$, $m(2^k) = 3 \cdot 2^{k-2}$
2. $m(5^k) = 5^k$
3. If p is an odd prime, $m(p^k) = m^{k-1} m(p)$
4. If p is a prime $10n \pm 1$, then $m(p)$ divides $p - 1$
5. If q is a prime $10n \pm 3$, then $m(q)$ divides $q + 1$

Furthermore, if $(r, s) = 1$, then $m(rs) = \dfrac{m(r)m(s)}{(m(r), \ m(s))}$

We now consider the primitive factors. For the equation $m(x) = x$, there are infinitely many solutions. They are all given by

$$x = 1, 5, 25, 125, 625, \ldots, 5^k, \ k = 1, 2, 3, \ldots$$
$$x = 12, 60, 300, 1500, 7500, \ldots, 12 \cdot 5^k, \ k = 0, 1, 2, 3, \ldots$$

We call these solutions of the equation $m(x) = x$, the primitive numbers.

Consider the sequence $m(n)$, $m(m(n))$, $m(m(m(n))) \ldots$. If we let n be any positive integer, then the sequence $m(n)$, $m(m(n))$,

$m(m(m(n)))$ . . . terminates in a primitive number, the second remarkable property of the Fibonacci numbers.

If one represents this function of the minor modulus $m(n)$ graphically, astonishing pictures arise.

So let n be a (given arbitrary) number. Then the smallest index of $x \in \mathbf{N}$ with $n \mid F_x$ (where F_x is a Fibonacci number) is said to be the *minor modulus* $m(n)$ of n.

At first, table 1 seems quite "chaotic" and also in its graphic presentation (figure 1).

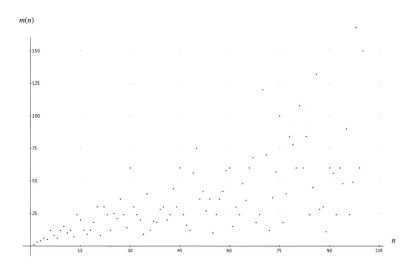

Figure 1

It rather looks like "fly droppings"—concentrated in the lower half of the quadrant, which lies under this one of diagonals going from the lower left to the upper right. If the points are connected (see figure 2), we recognize hardly any pattern.

$m(n)$

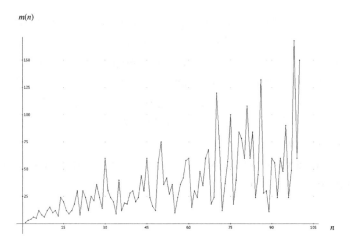

Figure 2

If we were to continue table 1 to about $n = 3,200$, then radial parts can be recognized in the graph—truly an amazing result, which gives us the third remarkable property of the Fibonacci numbers (see figure 3).

$m(n)$

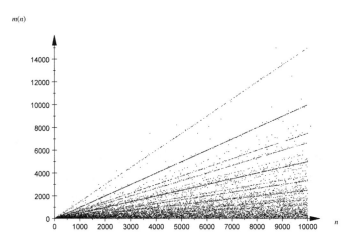

Figure 3

So you can see, even if a pattern does not immediately emerge, you may have to look a bit further until one appears. There are boundless beauties embedded in the Fibonacci numbers. I leave it to the reader to discover others.

The Ultimate Word

In this afterword we have discussed some of the divisibility properties of the Fibonacci numbers, and to this end have introduced the concept of the minor modulus $m(n)$. This work naturally raises the question: What can we say about those Fibonacci numbers that, upon division by the specified number n, leave the fixed remainder $r \neq 0$? Readers are invited to explore this question for themselves and to make some of the exciting discoveries that await them. The work is not easy but will be well rewarded.

As I read this delightfully informative book, I was motivated to search for new mathematical relationships regarding the Fibonacci numbers, some of which I just presented. I felt my search was incomplete until I could relate these ubiquitous numbers to the field that has occupied a large portion of my professional life—crystallography—and for which some of my work was acknowledged by the Nobel prize. My search turned up the following: a crystal may be defined mathematically as a triply periodic array of (or a triply periodic function, the electron density function) points (atoms). As such, it may have certain elements of symmetry, for example, a center of symmetry, a mirror plane, a two-fold rotation axis, and so on. Certain elements of symmetry, consistent with its triply periodic structure, are, on the other hand, forbidden—a simple example of forbidden symmetry is the five-fold symmetry element.

When, nevertheless, a crystal, the aluminum-manganese alloy, Al_6Mn, was observed in 1984 to have the forbidden fivefold symmetry, the crystallographers were presented with a dilemma. The dilemma was resolved with the recognition that this alloy constitutes a new state of matter, since called a quasi-crystal, having some properties of crystals and others of noncrystalline matter, such as glass.

Quasi-crystals may be described by means of a quasi-lattice; for example, in the one-dimensional case, by the Fibonacci lattice generated by the two basis vectors of lengths a and b defined by successive application of the rule $S_{n+1} = S_{n-1}S_n$, $n = 1, 2, 3 \ldots$, where

$$S_0 = a, \; S_1 = b, \; a = \phi b, \; \text{where} \; \phi = \frac{\sqrt{5}+1}{2} = 1.618034 \ldots$$

Then, as is readily confirmed,

$$S_0 = a, S_1 = b, S_2 = ab, S_3 = ab^2, S_4 = a^2 b^3, S_5 = a^3 b^5, S_6 = a^5 b^8, \ldots$$

In the Fibonacci lattice, the relationship of which to the sequence of Fibonacci numbers is clear, we have the fourth remarkable property of the Fibonacci numbers.

This example illustrates, once again, how often, and in what unexpected ways, the most abstract mathematical construct may throw light on a real-world phenomenon.

For some readers, my musings in the realm of the Fibonacci numbers may be a bit challenging. Yet I hope that each reader will see this fine book as a springboard to further investigations into the Fibonacci numbers and perhaps their relationship to the reader's field of interest. You have here a source for boundless entertainment and an entrée into the beautiful world of mathematics.

Appendix A

List of the First 500 Fibonacci Numbers, with the First 200 Fibonacci Numbers Factored

n	F_n	Number of digits	Factors for the first 200
1	1	(1)	unit
2	1	(1)	unit
3	2	(1)	prime
4	3	(1)	prime
5	5	(1)	prime
6	8	(1)	2^3
7	13	(2)	prime
8	21	(2)	$3 \cdot 7$
9	34	(2)	$2 \cdot 17$
10	55	(2)	$5 \cdot 11$
11	89	(2)	prime
12	144	(3)	$2^4 \cdot 3^2$
13	233	(3)	prime
14	377	(3)	$13 \cdot 29$
15	610	(3)	$2 \cdot 5 \cdot 61$
16	987	(3)	$3 \cdot 7 \cdot 47$
17	1597	(4)	prime
18	2584	(4)	$2^3 \cdot 17 \cdot 19$
19	4181	(4)	$37 \cdot 113$
20	6765	(4)	$3 \cdot 5 \cdot 11 \cdot 41$
21	10946	(5)	$2 \cdot 13 \cdot 421$
22	17711	(5)	$89 \cdot 199$
23	28657	(5)	prime
24	46368	(5)	$2^5 \cdot 3^2 \cdot 7 \cdot 23$
25	75025	(5)	$5^2 \cdot 3001$
26	121393	(6)	$233 \cdot 521$
27	196418	(6)	$2 \cdot 17 \cdot 53 \cdot 109$
28	317811	(6)	$3 \cdot 13 \cdot 29 \cdot 281$
29	514229	(6)	prime
30	832040	(6)	$2^3 \cdot 5 \cdot 11 \cdot 31 \cdot 61$
31	1346269	(7)	$557 \cdot 2417$
32	2178309	(7)	$3 \cdot 7 \cdot 47 \cdot 2207$
33	3524578	(7)	$2 \cdot 89 \cdot 19801$
34	5702887	(7)	$1597 \cdot 3571$
35	9227465	(7)	$5 \cdot 13 \cdot 141961$
36	14930352	(8)	$2^4 \cdot 3^3 \cdot 17 \cdot 19 \cdot 107$
37	24157817	(8)	$73 \cdot 149 \cdot 2221$
38	39088169	(8)	$37 \cdot 113 \cdot 9349$
39	63245986	(8)	$2 \cdot 233 \cdot 135721$
40	102334155	(9)	$3 \cdot 5 \cdot 7 \cdot 11 \cdot 41 \cdot 2161$
41	165580141	(9)	$2789 \cdot 59369$
42	267914296	(9)	$2^3 \cdot 13 \cdot 29 \cdot 211 \cdot 421$
43	433494437	(9)	prime
44	701408733	(9)	$3 \cdot 43 \cdot 89 \cdot 199 \cdot 307$
45	1134903170	(10)	$2 \cdot 5 \cdot 17 \cdot 61 \cdot 109441$
46	1836311903	(10)	$139 \cdot 461 \cdot 28657$
47	2971215073	(10)	prime
48	4807526976	(10)	$2^5 \cdot 3^2 \cdot 7 \cdot 23 \cdot 47 \cdot 1103$
49	7778742049	(10)	$13 \cdot 97 \cdot 6168709$
50	12586269025	(11)	$5^2 \cdot 11 \cdot 101 \cdot 151 \cdot 3001$
51	20365011074	(11)	$2 \cdot 1597 \cdot 6376021$
52	32951280099	(11)	$3 \cdot 233 \cdot 521 \cdot 90481$
53	53316291173	(11)	$953 \cdot 55945741$

n	F_n	Number of digits	Factors for the first 200
54	86267571272	(11)	$2^3 \cdot 17 \cdot 19 \cdot 53 \cdot 109 \cdot 5779$
55	139583862445	(12)	$5 \cdot 89 \cdot 661 \cdot 474541$
56	225851433717	(12)	$3 \cdot 7^2 \cdot 13 \cdot 29 \cdot 281 \cdot 14503$
57	365435296162	(12)	$2 \cdot 37 \cdot 113 \cdot 797 \cdot 54833$
58	591286729879	(12)	$59 \cdot 19489 \cdot 514229$
59	956722026041	(12)	$353 \cdot 2710260697$
60	1548008755920	(13)	$2^4 \cdot 3^2 \cdot 5 \cdot 11 \cdot 31 \cdot 41 \cdot 61 \cdot 2521$
61	2504730781961	(13)	$4513 \cdot 555003497$
62	4052739537881	(13)	$557 \cdot 2417 \cdot 3010349$
63	6557470319842	(13)	$2 \cdot 13 \cdot 17 \cdot 421 \cdot 35239681$
64	10610209857723	(14)	$3 \cdot 7 \cdot 47 \cdot 1087 \cdot 2207 \cdot 4481$
65	17167680177565	(14)	$5 \cdot 233 \cdot 14736206161$
66	27777890035288	(14)	$2^3 \cdot 89 \cdot 199 \cdot 9901 \cdot 19801$
67	44945570212853	(14)	$269 \cdot 116849 \cdot 1429913$
68	72723460248141	(14)	$3 \cdot 67 \cdot 1597 \cdot 3571 \cdot 63443$
69	117669030460994	(15)	$2 \cdot 137 \cdot 829 \cdot 18077 \cdot 28657$
70	190392490709135	(15)	$5 \cdot 11 \cdot 13 \cdot 29 \cdot 71 \cdot 911 \cdot 141961$
71	308061521170129	(15)	$6673 \cdot 46165371073$
72	498454011879264	(15)	$2^5 \cdot 3^3 \cdot 7 \cdot 17 \cdot 19 \cdot 23 \cdot 107 \cdot 103681$
73	806515533049393	(15)	$9375829 \cdot 86020717$
74	1304969544928657	(16)	$73 \cdot 149 \cdot 2221 \cdot 54018521$
75	2111485077978050	(16)	$2 \cdot 5^2 \cdot 61 \cdot 3001 \cdot 230686501$
76	3416454622906707	(16)	$3 \cdot 37 \cdot 113 \cdot 9349 \cdot 29134601$
77	5527939700884757	(16)	$13 \cdot 89 \cdot 988681 \cdot 4832521$
78	8944394323791464	(16)	$2^3 \cdot 79 \cdot 233 \cdot 521 \cdot 859 \cdot 135721$
79	14472334024676221	(17)	$157 \cdot 92180471494753$
80	23416728348467685	(17)	$3 \cdot 5 \cdot 7 \cdot 11 \cdot 41 \cdot 47 \cdot 1601 \cdot 2161 \cdot 3041$
81	37889062373143906	(17)	$2 \cdot 17 \cdot 53 \cdot 109 \cdot 2269 \cdot 4373 \cdot 19441$
82	61305790721611591	(17)	$2789 \cdot 59369 \cdot 370248451$
83	99194853094755497	(17)	prime
84	160500643816367088	(18)	$2^4 \cdot 3^2 \cdot 13 \cdot 29 \cdot 83 \cdot 211 \cdot 281 \cdot 421 \cdot 1427$
85	259695496911122585	(18)	$5 \cdot 1597 \cdot 9521 \cdot 3415914041$
86	420196140727489673	(18)	$6709 \cdot 144481 \cdot 433494437$
87	679891637638612258	(18)	$2 \cdot 173 \cdot 514229 \cdot 3821263937$
88	1100087778366101931	(19)	$3 \cdot 7 \cdot 43 \cdot 89 \cdot 199 \cdot 263 \cdot 307 \cdot 881 \cdot 967$
89	1779979416004714189	(19)	$1069 \cdot 1665088321800481$
90	2880067194370816120	(19)	$2^3 \cdot 5 \cdot 11 \cdot 17 \cdot 19 \cdot 31 \cdot 61 \cdot 181 \cdot 541 \cdot 109441$
91	4660046610375530309	(19)	$13^2 \cdot 233 \cdot 741469 \cdot 159607993$
92	7540113804746346429	(19)	$3 \cdot 139 \cdot 461 \cdot 4969 \cdot 28657 \cdot 275449$
93	12200160415121876738	(20)	$2 \cdot 557 \cdot 2417 \cdot 4531100550901$
94	19740274219868223167	(20)	$2971215073 \cdot 6643838879$
95	31940434634990099905	(20)	$5 \cdot 37 \cdot 113 \cdot 761 \cdot 29641 \cdot 67735001$
96	51680708854858323072	(20)	$2^7 \cdot 3^2 \cdot 7 \cdot 23 \cdot 47 \cdot 769 \cdot 1103 \cdot 2207 \cdot 3167$
97	83621143489848422977	(20)	$193 \cdot 389 \cdot 3084989 \cdot 361040209$
98	135301852344706746049	(21)	$13 \cdot 29 \cdot 97 \cdot 6168709 \cdot 599786069$
99	218922995834555169026	(21)	$2 \cdot 17 \cdot 89 \cdot 197 \cdot 19801 \cdot 18546805133$
100	354224848179261915075	(21)	$3 \cdot 5^2 \cdot 11 \cdot 41 \cdot 101 \cdot 151 \cdot 401 \cdot 3001 \cdot 570601$
101	573147844013817084101	(21)	$743519377 \cdot 770857978613$
102	927372692193078999176	(21)	$2^3 \cdot 919 \cdot 1597 \cdot 3469 \cdot 3571 \cdot 6376021$
103	1500520536206896083277	(22)	$519121 \cdot 5644193 \cdot 512119709$
104	2427893228399975082453	(22)	$3 \cdot 7 \cdot 103 \cdot 233 \cdot 521 \cdot 90481 \cdot 102193207$
105	3928413764606871165730	(22)	$2 \cdot 5 \cdot 13 \cdot 61 \cdot 421 \cdot 141961 \cdot 8288823481$
106	6356306993006846248183	(22)	$953 \cdot 55945741 \cdot 119218851371$
107	10284720757613717413913	(23)	$1247833 \cdot 824206505006l761$
108	16641027750620563662096	(23)	$2^4 \cdot 3^4 \cdot 17 \cdot 19 \cdot 53 \cdot 107 \cdot 109 \cdot 5779 \cdot 11128427$
109	26925748508234281076009	(23)	$827728777 \cdot 32529675488417$
110	43566776258854844738105	(23)	$5 \cdot 11^2 \cdot 89 \cdot 199 \cdot 331 \cdot 661 \cdot 39161 \cdot 474541$
111	70492524767089125814114	(23)	$2 \cdot 73 \cdot 149 \cdot 2221 \cdot 1459000305513721$
112	114059301025943970552219	(24)	$3 \cdot 7^2 \cdot 13 \cdot 29 \cdot 47 \cdot 281 \cdot 14503 \cdot 10745088481$
113	184551825793033096366333	(24)	$677 \cdot 272602401466814027129$
114	298611126818977066918552	(24)	$2^3 \cdot 37 \cdot 113 \cdot 229 \cdot 797 \cdot 9349 \cdot 54833 \cdot 95419$
115	483162952612010163284885	(24)	$5 \cdot 1381 \cdot 28657 \cdot 2441738887963981$
116	781774079430987230203437	(24)	$3 \cdot 59 \cdot 347 \cdot 19489 \cdot 514229 \cdot 1270083883$
117	1264937032042997393488322	(25)	$2 \cdot 17 \cdot 233 \cdot 29717 \cdot 135721 \cdot 39589685693$
118	2046711111147398462369 1759	(25)	$353 \cdot 709 \cdot 8969 \cdot 336419 \cdot 2710260697$
119	3311648143516982017180081	(25)	$13 \cdot 1597 \cdot 159512939815855788121$
120	5358359254990096664087 1840	(25)	$2 \cdot 5 \cdot 3^2 \cdot 5 \cdot 7 \cdot 11 \cdot 23 \cdot 31 \cdot 41 \cdot 61 \cdot 241 \cdot 2161 \cdot 2521 \cdot 20641$
121	8670007398507948658051921	(25)	$89 \cdot 97415813466381445596089$
122	14028366653498915298923761	(26)	$4513 \cdot 555003497 \cdot 5600748293801$
123	22698374052006863956975682	(26)	$2 \cdot 2789 \cdot 59369 \cdot 6854195773 3949701$
124	36726740705505779255899443	(26)	$3 \cdot 557 \cdot 2417 \cdot 3010349 \cdot 3020733700601$
125	59425114757512643212875125	(26)	$5^3 \cdot 3001 \cdot 158414167964045700001$
126	96151855463018422468774568	(26)	$2^3 \cdot 13 \cdot 17 \cdot 19 \cdot 29 \cdot 211 \cdot 421 \cdot 1009 \cdot 31249 \cdot 35239681$
127	155576970220531065681649693	(27)	$27941 \cdot 5568053048227732210073$
128	251728825683549488150424261	(27)	$3 \cdot 7 \cdot 47 \cdot 127 \cdot 1087 \cdot 2207 \cdot 4481 \cdot 186812208641$
129	407305795904080553832073954	(27)	$2 \cdot 257 \cdot 5417 \cdot 8513 \cdot 39639893 \cdot 433494437$
130	659034621587630041982498215	(27)	$5 \cdot 11 \cdot 131 \cdot 233 \cdot 521 \cdot 2081 \cdot 24571 \cdot 14736206161$
131	1066340417491710595814572169	(28)	prime
132	1725375039079340637797070384	(28)	$2^4 \cdot 3^2 \cdot 43 \cdot 89 \cdot 199 \cdot 307 \cdot 9901 \cdot 19801 \cdot 261399601$
133	2791715456571051233611642553	(28)	$13 \cdot 37 \cdot 113 \cdot 3457 \cdot 42293 \cdot 351301301942501$
134	4517090495650391871408712937	(28)	$269 \cdot 4021 \cdot 116849 \cdot 1429913 \cdot 24994118449$
135	7308805952221443105020355490	(28)	$2 \cdot 5 \cdot 17 \cdot 53 \cdot 61 \cdot 109 \cdot 109441 \cdot 1114769954367361$
136	11825896447871834976429068427	(29)	$3 \cdot 67 \cdot 1597 \cdot 3571 \cdot 63443 \cdot 23230657239121$
137	19134702400093327808144942 3917	(29)	prime
138	30960598847965113057878492344	(29)	$2^3 \cdot 137 \cdot 139 \cdot 461 \cdot 691 \cdot 829 \cdot 18077 \cdot 28657 \cdot 1485571$
139	50095301248058391139327916261	(29)	$277 \cdot 2114537501 \cdot 8552672293768 9093$
140	81055900096023504197206408605	(29)	$3 \cdot 5 \cdot 11 \cdot 13 \cdot 29 \cdot 41 \cdot 71 \cdot 281 \cdot 911 \cdot 141961 \cdot 12317523121$
141	131151201344081895336534324866	(30)	$2 \cdot 108289 \cdot 1435097 \cdot 142017737 \cdot 2971215073$
142	212207101440105399533740733471	(30)	$6673 \cdot 46165371073 \cdot 688846502588399$
143	343358302784187294870275058337	(30)	$89 \cdot 233 \cdot 8581 \cdot 1929584153 75685 0496621$
144	555565404224292694404015791808	(30)	$2^5 \cdot 3^3 \cdot 7 \cdot 17 \cdot 19 \cdot 23 \cdot 47 \cdot 107 \cdot 1103 \cdot 103681 \cdot 10749957121$
145	898923707008479989274290850145	(30)	$5 \cdot 514229 \cdot 34961999693073 7079890201$
146	1454489711123277263367830641953	(31)	$151549 \cdot 9375829 \cdot 86020717 \cdot 11899937029$
147	2353412818241252672952597492098	(31)	$2 \cdot 13 \cdot 97 \cdot 293 \cdot 421 \cdot 3529 \cdot 6168709 \cdot 347502052673$
148	3807901929474025356630904134051	(31)	$3 \cdot 73 \cdot 149 \cdot 2221 \cdot 11987 \cdot 54018521 \cdot 81143477963$
149	6161314747715278029583501626149	(31)	$110557 \cdot 162709 \cdot 4000949 \cdot 85607646594577$
150	9969216677189303386214405760200	(31)	$2^3 \cdot 5^2 \cdot 11 \cdot 31 \cdot 61 \cdot 101 \cdot 151 \cdot 3001 \cdot 12301 \cdot 18451 \cdot 230686501$
151	16130531424904581415797907386349	(32)	$5737 \cdot 2811666624525811664 69915877$

n	F_n	Number of digits	Factors for the first 200
152	260997481020938848020123131465449	(32)	3·7·37·113·9349·29134601·1091346396980401
153	422302795269984466217810220532898	(32)	2·17²·1597·6376021·7175323114950564593
154	683300276290923510198225336794447	(32)	13·29·89·199·229769·988681·4832521·9321929
155	1105603071560908172376327542122345	(33)	5·557·2417·21701·12370533881·61182778621
156	1788903347851831682574552878917924	(33)	2⁴·3²·79·233·521·859·90481·135721·12280217041
157	2894506419412739854950880421041137	(33)	313·11617·7636481·1042420430649134 6737
158	4683409767264571537525433299954929	(33)	157·92180471494753·3236112267 2259149
159	7577916186677311392476313721000066	(33)	2·317·953·55945741·97639037·229602768949
160	12261325953941882930001747020959 95	(34)	3·5·7·11·41·47·1601·2161·2207·3041·23725145626561
161	19839242140619194322478060741960 61	(34)	13·8693·28657·6126061077550589970655 97
162	32100568094561077252479807769205 6	(34)	2³·17·19·53·109·2269·3079·4373·5779·19441·62650261
163	51939810235180271574957868504881 17	(34)	977·4892609·33365519393·32566223208133
164	84040378329741348827437676267801 73	(34)	3·163·2789·59369·800483·350207569·370248451
165	13598018856492162040239554477268290	(35)	2·5·61·89·661·19801·86461·474541·518101·900241
166	22002056689466296922983322104048463	(35)	35761381·6202401259·99194853094755497
167	35600075545958458963222876581316753	(35)	18104700793·1966344318693345608565721
168	57602132235424755886206198685365216	(35)	25·3²·7²·13·23·29·83·167·211·281·421·1427·14503·65740583
169	93202207781383214849429075266681969	(35)	233·337·89909·104600155609·126213229732669
170	150804340016807970735635273952047185	(36)	5·11·1597·3571·9521·1158551·12760031·3415914041
171	244006547798191185585064349218729154	(36)	2·17·37·113·797·6841·54833·574146176087944 361
172	394810887814999156320699623170776339	(36)	3·6709·144481·433494437·31319571151657 8281
173	638817435613190341905763972389505493	(36)	1639343785721·3896787490007629271532733
174	1033628323428189498226463595560281832	(37)	2³·59·173·349·19489·514229·947104099·3821263937
175	1672445759041379840132227567949787325	(37)	5²·13·701·3001·141961·17231203730201189308301
176	2706074082469569338358691163510069157	(37)	3·7·43·47·89·199·263·307·881·967·93058241·562418561
177	4378519841510949178490918731459854482	(37)	2·353·2191261·805134061·1297027681·2710260697
178	7084593923980518516849609894969925639	(37)	179·1069·1665088321800481·22235502640988369
179	11463113765491467695340528624629782121	(38)	21481·156089·341881664090389929534613769
180	18547707689471986212190138521399707760	(38)	2⁴·3³·5·11·19·31·41·61·107·181·541·2521·109441·10783342081
181	30010821454963453907530667147829489881	(38)	8689·422453·81757892372385475745514 61093
182	48558529144453440119720805669229 19641	(38)	13²·29·233·521·741469·13960793·689667151970161
183	78569350599398894027251472817058687522	(38)	2·1097·4513·555003497·142973479719757 5800833
184	127127879743834334146972278486287885163	(39)	3·7·139·461·4969·28657·253367·275449·9506372193863
185	205697230343233228174223751303346572685	(39)	5·73·149·2221·1702945513191305556907097618161
186	332825110087067562321196029789634457848	(39)	2⁵·557·2417·63799·3010349·35510749·4531100550901
187	538522340430300790495419781092981030533	(39)	89·373·1597·10157807305963434099105034917037
188	871347450517368352816615810828251488381	(39)	3·563·5641·2971215073·661·2971215073·4632894751907
189	1409869790947669143312035591975965189 14	(40)	2·13·17·53·109·421·38933·35239681·95592195031 6735037
190	2281217241465037496128651402858212007295	(40)	5·11·37·113·191·761·9349·29641·41611·67735001·87382901
191	3691087032412706639440686994833808526209	(40)	4870723671313·7578108062569891284399 75793
192	5972304273877744135569338397692020533504	(40)	2³·3²·7·23·47·769·1087·1103·2207·3167·4481·11862575248703
193	9663391306290450775010025392525829059713	(40)	9465278929·102093043203232693397682 6008497
194	15635695580168194910579363790217849593217	(41)	193·389·3299·3084989·361040209·56678557502141579
195	25299086886458645685589389182743678652930	(41)	2·5·61·233·135721·14736206161·8899925083749987 7681
196	40934782466626840596168752972961528246147	(41)	3·13·29·97·281·5881·49033·148739·61025309469041
197	66233869353085486281758142155705206899077	(41)	15761·25795969·227150265697·7171851 07125886549
198	107168651819712326877926895128666735145224	(42)	2³·17·19·89·197·199·991·2791·19801·1513909·18546805133
199	173402521172797813159685037284371942044301	(42)	397·4367821692010020482611713785500 55269633
200	280571172992510140037611932413038677189525	(42)	3·5²·7·11·41·101·151·401·2161·3001·570601·9125201

201	453973694165307953197296969697410619233826
202	734544867157818093234089021104492964233 51
203	1188518561323126046432205871807859915657177
204	1923063428480944139667114773918309212080528
205	3111581989804070186099320645726169127737705
206	5034645418285014325766435419644478339818233
207	8146227408089084511865756065370647467555938
208	13180872826374098837632191485015125807374171
209	21327100234463183349497947550385773274930109
210	34507973060837282187130139035400899082304280
211	55835073295300465536628086585786672357234389
212	90343046356137747723758225621187571439538669
213	146178119651438213260386312206974243796773058
214	236521166007575960984144537828161815236311727
215	382699285659014174244530850035136059033084785
216	619220451666590135228675387863297874269396512
217	1001919737325604309473206237898433933302481297
218	1621140188992194444701881625761731807571877809
219	2623059926317798754175087863660165740874359106
220	4244200115309993198876969489421897548446236915
221	6867260041627791953052057353082063289320596021
222	11111460156937785151929026842503960837766832936
223	17978720198565577104981084195586024127087428957
224	29090180355503362256910111038089984964854261893
225	47068900554068939361891195233676009091941690480
226	76159080909057230161880130627176599405679595 2743
227	123227981463641240980692501505442003148737643593
228	199387062373213542599493807777207997205533596336
229	322615043836854783580186309282650000354271239929
230	522002106210068326179680117059857997559804836265
231	844617150046923109759866426342507997914076076194
232	1366619256256991435939546543402365995473880912459
233	2211236406303914545699941296974487399338795698865 3
234	3577855662560905981638959513147239988861837901112
235	5789092068864820527338372428929113982249794889765
236	9366947731425726508977331996039535397111632790877
237	15156039800290547036315700447893146795336142768064 2
238	24522987531716273545290364749708219244730604715 19
239	3967902733200682058160874095390228987783448815216 1
240	6420201486372309412690177742887311802307548623680
241	10388104219572991470851051838277540168014203 6775841
242	16808305705945300883541229581164851343422 4495395521
243	27196409925518292354392281419442391516259162 2175362
244	44004715631463593237933511000607242864504 1207574883
245	71201125556981885592325792420049634380 7632829750245

n	F_n
246	1152058411884454788302593034206568772452674037325128
247	1864069667454273644225850958407065116260306867075373
248	3016128079338728432528443992613633888712980904400501
249	4880197746793002076754294951020699004973287771475874
250	7896325826131730509282738943634332893686268675876375
251	12776523572924732586037033894655031898659556447352249
252	20672849399056463095319772838289364792345825123228624
253	33449372971981195681356806732944396691005381570580873
254	54122222371037658776676579571233761483351206693809497
255	87571595343018854458033386304178158174356588264390370
256	141693817714056513234709965875411919657707794958199867
257	229265413057075367692743352179590077832064383222590237
258	370959230771131880927453318055001997489772178180790104
259	600224643828207248620196670234592075321836561403380341
260	971183874599339129547649988289594072811608739584170445
261	1571408518427546378167846658524186148133445300987550786
262	2542592393026885507715496646813780220945054040571721231
263	4114000911454431885883343305337966369078499341559272017
264	6656593304481317393598839952151746590023553382130993248
265	10770594215935749279482183257489712959102052723690265265
266	17427187520417066673081023209641459549125606105821258513
267	28197781736352815952563206467131172508227658829511523778
268	45624969256769882625644229676772632057353264935332782291
269	73822750993122698578207436143903804565580923764844306069
270	119447720249892581203851665820676436622934188700177088360
271	193270471243015279782059101964580241188515112465021394429
272	312718191492907860985910767785256677811449301165198482789
273	505988662735923140767969869749836918999964413635240577218
274	818706854228831001753880637535093596811413714795418360007
275	1324695516964754142521850507284930515811378128425638237225
276	2143402371193585144275731144820024112622791843221056597232
277	3468097888158339286797581652104954628434169971646694834457
278	5611500259351924431073312796924978741056961814867751431689
279	9079598147510263717870894449029933369491131786514446266146
280	14691098406862188148944207245954912110548093601382197649735
281	23770696554372451866815101694984845480039225387896643963981
282	38461794961234640015759308940939757590587319989278846161816
283	62232491515607091882574410635924603070626544377175485625797
284	100694286476841731898333719576864360661213863366454327287613
285	162926777992448823780908130212788963731840407743629812913410
286	263621064466290555679241849789653324393054271110084140201023
287	426547842461739379460149980002442288124894678853713953114433
288	690168906931029935133939182979209561251794849964963798093315456
289	1116716749392769314599541809794537900642843628817512046429889
290	1806885656323799249738933639586633513160792578781310139745345
291	2923602405716568564338475449381171413803636207598822186175234
292	4730488062040367814077409089678049269644287863801323325920579
293	7654090467756936378415884538348976340768064993978954512095813
294	12384578529797304192493293627316781267732493780359086838016392
295	20038668997554240570909178165665757608500558774330841350112205
296	32423247527351544763402471792982538876233052554697128188128597
297	52461916524905785334311649958648273361132903516953824080 2
298	84885164052257330097714121751630835360966663883732297726369399
299	137347080577163115432025771710279131845700275212767467264610201
300	222232244629420445529739893461909967206666939096499764990979600
301	359579325206583560961765665172189099052367214309267232255589801
302	581811569836004006491505558634099066259034153405766099724659 4 401
303	941390895042587567453271223806288165311401367715034229502159202
304	1523202464878591573944776782440387217504355211208012267487286 0 3
305	2464593359921179141398048006246675396881836888358535456250887805
306	3987795824799770715342824788637806262845227240995663668299961640 8
307	6452389184720949856740872794933738025334109298792472139250504213
308	10440185009520720572083697583620806537863817078470108822250120621
309	16892574194241670428824570378554538679120491007541580961500624834
310	27332759203762391000908267962175339332906872716290689783750745455
311	44225333398004061429732838340729878012027363723832270745251370289
312	71558092601766452430641106302905217344934236440122960529002115744
313	115783425999770513860373944463635095356961600163955523127425486033
314	187341518601536966291015050946540312701895836604078191803255601777
315	303124944601307480151388995590175408058857436788343423077509087810
316	490466463202844446442404046536715720760753273372111614880764689587
317	793591407804151926593793042126891128819610710140145037958273977397
318	1284057871006996373036197088663606849580363983512256652839038466984
319	2077649278811148299629990130790496797469365241690797312244381
320	3361707149818144672666187219454104827980338677164658343636350711365
321	5439356428629292972296177350244602806380313370817060034433662955746
322	8801063578447437644962364569698707634360652047981718378070013667111
323	14240420007076730617258541919943310440740965418798778412503676622857
324	23041483585524168262220906489642018075101617466780496790573690289968
325	37281903592600898879479448409585328515842582885579275203077366912825
326	60323387178125067141700354899227346590944200352359771993651057202793
327	97605290770725966021179803308812675106786783237939047196728424115618
328	157928677948851033162880158208040021697730983590298819190379481318411
329	255533968719576999184059961516852696804517766828237866387107905434029
330	413462646668428032346940119724892718502248750418536857748738675 2440
331	668996615388005031531000081241745415306766517246774551964595292186469
332	1082459262056433063877940200966638133809015267665311237542082678939
333	1751455877444438095408940282200838354915178178491208578950667797112537 8
334	2833915139500871159286880483175021682924797052577397027048760650064287
335	4585371016945309254695820765383405278837489482816555438621189665
336	7419286156446180413982701248558426914965375890066879843604199271253952
337	12004657173391489668678522013941832147005954727556326601596378924443611
338	19423943329837670082661223262500259061971330617623242503763837163697569
339	31428600503229159751339745276442091208977285345179605163923475056141186
340	50852543833066829834000968538942350270948615962802847667687312219838755

n	F_n
341	82281144336295989585340713815384441479925901307982452831610787275979941
342	133133688169362819419341682354326791750874517270785300499298099495818696
343	215414832505658809004682396169711233230800418578767753330908886771798637
344	34854852067502162842402407852403802498167493584955305383020698626717333
345	563963353180680437428706474693749258212475354428320807161115873039415970
346	9125118738557020658527305532177872831941502902778738600991322859307033303
347	1476475227036382503281437027911536541406625644706194668152438732346449273
348	2388987100892084569134167581129323824600775934984068529143761591653482576
349	3865462327928467072415604609040860366007401579690263197296200323999931849
350	6254449428820551641549772190170184190608177514674331726439961915653414425
351	10119911756749018713965376799211044556615579094364594923736162239653346274
352	16374361185569570355515148989381228747223756609038926650176124155306760699
353	264942729423185890694805257885922733038393353703403521573912286394960106973
354	42868634127888159424995674777973502051063092312442448224088410550266867672
355	69362907070206748494476200566565775354902428015845969798000696945226974645
356	112231541198094907919471875344539277405965520328288418022089107495493842317
357	181594448268301656413948075911105052760867948344134387820089804440720816962
358	293825989466396564333419951255644330166833468672422805842178911936214659279
359	475420437734698220747368027166749382927701417016557193662268716376935476241
360	769246427201094785080787978422393713094534885688979999504447628313150135520
361	1244666864935793005828156005589143096022236302705537193166716344690085611761
362	2013913292136887790908943984011536809116771188394517192671163973003235747281
363	32585880157072680796737099989600679905139007491100054385837880317693321359042
364	5272493449209568587646043972612167142557867949457151739122863949600657106323
365	8531073606282249384383143963212896619394786170594625964346924608389878465365
366	13803567055491817972029187936825113333650564850089197542855968990806435571688
367	22334640661774067356412331900038009953045351020683823507202893507476314037053
368	36138207717265885328441519836863123286606959158707730210500588624065627496087 41
369	584728483790399526848538517369011332397412668914568444557261755914039063645794
370	9461105609630583801329537152737642565626437182762229865607320618320601813254535
371	153083904475345790698149223310665389766178449653686710164582374234640876900329
372	2476949460571651628711444594884429646292615632415916575771902992555242690154864
373	400778865046999741940959381819509503657894082069603285936485366789883567055193
374	64847382561864904812103841307952468235140971448551986170838359345126257210057
375	1049252690665646467530632231274619718410203796555123147644873726135009824265250
376	1697726516284295515651670644354144400761613511040643009353262085480136081475307
377	274697920694994419831823028756287641191718173075957640755444011615145905740557
378	4444705723234237498833973519982908519933430818636409166351397897095281987215864
379	7191684930184179482016726395611672639105248126232175323349533708710427892956421
380	11636390653418416980850249915594581159038678944868584489700931605805709880172285
381	18828075583602596462866526311206253798143927071100579813050465314516137773128706
382	30464466237021013443716776226800834957182606015969344302751396920321847653300991
383	49292544182062360990658330253800708875532653380701004115801862248379854262649697
384	79757008057644623350300078764807923712509139103039448418553259155159833079730688
385	129049549878268233256883381302815012467835672190109552534355121389997818506160385
386	208806557935912856607183460067622936180344811293149000952908380545157651585891073
387	337856107814181089864066841370437948648180483483258553487263501935155470092051458
388	5466626657500939464712503014380600884828525294776407554440171882480313121677942531
389	884518773564275036335317142808498833476705778259666107927435384415468591769993989
390	14311811439314368982806567444246559718305231078303507303608871173447936520
391	231570021287864401914188458705505855178193685129573977029504265131125030521793050951...
392	3746881652190313001948452031316182700871679243318134326626490918207032018665867029
393	606258186507165702109033661835667682186910477562755320295769256951828232388379538
394	98094635172646700230387886496582950919567269995936665362034248772534142436967599
395	15872045382336327044129125268014971913825377475586919838578035057243596666433462105
396	2568150889960099706716791391767326700578165017554628647419837544968911008983126672
397	415553542819373241112970391856882389196070276511332063127764126022125076754165887 77
398	67235063181538321178464953103361505925388677826679492786974790147181418868439971544 9
399	108788617463475645289761992289049744844995705477814292097439392635996163042 26
400	17602368064501396646822694539241125077038438330492191886725992896575345044216019675
401	284812298108489611757988937681460995615380088782034890986477195645969271404032323901
402	460835978753503578226215883073872246385764472086797082873203188542544616448248343576
403	74564827686199318998420482075533324200114456086910191735948544188513887523806671
404	1206484255615496768210420703829205488386909032955899056732883572731058504300529011053
405	1952132532477489958194625524584538730388053593825001030592563956919572392152809678530
406	315861678809298672640504622841374421877496262678090008732544752965063089964533368689583
407	5110749320570476684599671752998282949163016220605901117918011486570203288606148368113
408	8269366108663463411004717981412027167937789243181343426624311485059487057696
409	13380115429233940095604389734410310117100995067992702323161470502791037473665635425809
410	21649481537897403506609107715822337285038973915379205851566400021802909132390757909314
411	3502959696713134360221349745023264740213996898337220585156640001802909132390757909314
412	56679078505028747108822605166054984681717894289875170937997132954081478079111588392819
413	91708675472160090711036102616287632089318911882123915231537729562617689923506638302133
414	14838775397718883781985870778234261676497854788171509059103432470714622518694952
415	24009642944934892853089481039863024886581676666299953984304678866605016063812915 6997085
416	38848418342653776635075351818097286564231462375164458577694482631558472561675692037
417	628580612875886694816483285796031145081313810687470429760263643553279199088083268912 2
418	10170647963024244612324018467605759801504460095507498687521584842050154233436325083811 59
419	16456454091783111561140501753401790946585773976576263108974851680734562115334513341070281
420	26627102054807356173464520221007550748090234072083744411801919604845563638678145849451440
421	43083556144659046773460502179744093416946766267809000873254475296506308964533368689 583161
422	697106582013978239080695421954168924427662421207437345665360033033167549269080503997316 1
423	11279421434798829164267456416982621374422501694037247150528105518773677346703464230494882
424	18250487254938611555074410636524312658020849229014745928158881386149462839394269270468043
425	2952990686897374407193418670535069360717650742459532914624419672501860977335009 62925
426	477803959446760522744162776900312487297859234749698643278230343828116713025492002771430968
427	77310304634413492993758144743538184801550997720924982727487206270083963211589736272393893
428	125090700579089545268174422433569433531336921195894847055310250098200676237081739043824861
429	2024010052135030382619325671771076183328879189168198297827974563682846394486714753162187 54
430	32749170579259258530106989610677058462248401127146764684853156857532143600436154
431	52989271100609562179203955678778467019711275902953450662090516283476995513442468967626236 9
432	85738441679868820532214654639846172353614514314871538930901437280496774780497748874337
433	13872771278047838271114186103186246392258450358171783690079918032136025225954602593712568353
434	22446615446034720324363326495847081143197879573140328735389309014372804967747804977488743 37
435	363193867240825585955051875277095450657823831548581656361884893357330572272938309146144 2690

n	F_n
436	5876600217011727891986851402355662620898026272799849437157779835010586219504163589210317027
437	9508538889419983751537370155126617127476264588285666000776628768583891942233546680671759717
438	15385139106431711643524221557482279748374290861085515437934408603594478161737710269882076744
439	2489367799585169530615917126088968758505554493711814387110373721783701039712569505053836461
440	4027881710228340703858581327009117662422484631045669687664544597577248265708967220435913205
441	65172495098135102433647404982700073500075401759827878315356483347951218369680224170989749666
442	1054531312200418509472233218252791250124300248070284575192001929323724066635389191391425662871
443	170623807298553611905880623235491323624375649830112453507358412671675285005069415562415412537
444	276075119498972121378113841488282573748675897900397028699360341995399351640458606953841075408
445	446698926797525733283994464723773897373051547730509482206718754667074636645528022516256487945
446	722774046296497854662108306212056471121727445630906510906079096662473988285986629470097563353
447	1169472973094023587946102770935830368494778993361415993112797851329548624931514651986354051298
448	1892247019390521442608211077147886839616506438992322504018876947992022613217501281456451614651
449	3061719992484545030554313848083717208111285432353738497131674799321571238149015933442805665949
450	4953967011875066473162524925231604047727791871346061001150551747313593851366517214899257280600
451	801568700435961150371683773315321255839077303699799498282226546635165089515533148342062946549
452	12969654016234677976879363698546925303566869175045860499432778293948758940882050363241320227149
453	20985341020594289480596202471862246559405946478745659997715004840583924030397583511583383173698
454	339549950368289674574775566170409171862972815653791520497147783134532682971279633874824703400847
455	54940336057423256938071768642271418422378762132537180494862787975116607001677217386408086574545
456	88895331094252224395547334812668053515577786328700992010571109649289972956851261232789975392
457	143835667151675481333619103454952008707730339918865881486873359084765896974634068647640876549937
458	232730998245927705729166438267632598993081917705194582478883930194415186947590919908873666525329
459	376566665397603187062785541722584607700812257624060463965757289279181083922224988556514543075266
460	609297663643530892791951979990217206693894175329255046444641219473596270869815908465388209600595
461	985864329041134079854737521712801814394706432953315610410398508752777354792040897021902752675861
462	1595161992684664972646689501703019021088600608282570556855039728226373625661856805487290962276456
463	2581026321725799052501427023415820835483307041235886067265438236979150980453897702509193714952317
464	4176188314410464025148116525118839856571907649518456624120477965205524606115754507996484677228773
465	6757214636136263077649543548816681600711637324632164193714499386926652484692790595738232
466	10933402950546727102797660073653500548627122340272799315506394167390200192685406718502163069409863
467	176906117586682990180447203622188161240682337031027142006892310369574875779255058929007841461590953
468	28624020537229717283244863695841661789309459371299941322398704536965075971940465647510004531000816
469	46314638123912707463692067318029823029991796402327083329291014906539951751195524576517845945992591769
470	74938658661142424746936931013871484819301255773627024651689719443505027723135990224027850523592585
471	121253296785055132210628998331901307849293052175954107980980734350044979474331514800545696516184354
472	1961919554461975569575659293457727926628457793617681930113263267045379355000719746750502457354703977699639
473	317445252231252689168194927677674100517887360125535240613651188143594986671799019825119243555961293
474	513637207677450246125760855702344689318461668075116373246321641937144993869266524849692790595738232
475	831082459908702935293955784701120993704369028200651613859972830080739980541065544674812034151699525
476	1344719667586153181419716641724567886980506962757679871062944720178849744103320695245048248247437757
477	21758021274948561167136724264256888805952197244764196009662673020986249549513976141993658588991373282
478	35205217950810092981333890681502567674860704020752187588072561774116509929361729683723821683646575039
479	56963239225758654148470614945759456480812901452286071890388290762151348843131272979231385425417512321
480	9216845717656874712980450562726202415567360565980794777111390850331644813674856981646960226192287360
481	14913169640232740127827512057030214806364868765120940196615021992564677969798798427957009879687373999681
482	24130015357889614840807962620028350479216011277190196743261610776878424511662841261217058994930287041
483	390431849981223549686353456477330498543830704923181830703425204209650825408787157763668286722
484	6317320035601196980944343729735884902208067326558979545267344148030362872131366682004216758598573763
485	1022163853541343247780789119746893475649453352539893941620852728143783293096449232417345222666860485
486	165389585710146294587522349272048196587026008519579189614758713664032461652278159144795591280865434248
487	2676059710642806193656012612467375441519713437735685837768439858477612945832426514875869658031322947
488	43299555677442691395312361051878574073899735229314777339160269951179375623552081063238255708399772898
489	700601527838707533318724871765523284890968696066716357168446685359555050818763462119969522887130023714
490	11335570846131344472718484822843090256299660483598641305600493848713488070542844272752352079711277527695
491	18341986124518419805905733540498323105209347444265084087728496070230903857873047734872321602858257776409
492	2967795697064976427862421836334141336150900792786444618288545455102252664927332007624673682829385529104
493	48019943095168184084529951908383973646671835537213025106017041525333156522800379742496995285687643305513
494	77697900065817498363154170267181149828227363299999469724305586980435409187727711750121668968517028834617
495	12571784316098613244768412217102088629494571867212494830322628505768565710528091492618664254204672140130
496	2034157432268040808108382924382020935941068326060883324529903470149056115823592713458322272170097474747
497	32913358638779021325852241460922292241811880064424459384950843991972540608783894735358997476926373114877
498	5325493296145942940693607070474249585412918826163642393957905947817651550703969797809933069496480740896249624
499	861682916002384507327883121656647880959410683260608332452990340149056115823592713458328176574447204501
500	139423224561697880139724382870407283950070256587697307264108962948325571622863290691557658876222521294125

Appendix B

Proofs of Fibonacci Relationships

For Chapter 1

Some of the following proofs involve a technique called mathematical induction. A brief explanation of this procedure may be in order here.

Introduction to Mathematical Induction

Consider a set of dominoes with an endless number of tiles set up as illustrated below.

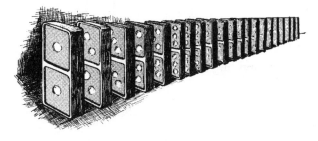

Figure B-1

If we were asked to knock down all the tiles, we could consider two methods:

(1) we could knock down each tile separately; or
(2) we could knock down just the first tile, if we were sure that any tile knocked down would automatically knock down the tile after it.

The first method would not only be inefficient but would also never assure us of knocking down all the tiles (since the end may never be reached). The second method would guarantee that all the tiles are knocked down. If we knocked down the first tile, we would be assured that it, and any knocked down tile, would knock down its successor tile—that is, the first tile knocks down the second, which knocks down the third, which knocks down the fourth, and so on. All would be knocked down.

This second method is directly analogous to the *axiom of mathematical induction*:

A proposition involving the natural number n is true for all natural numbers:

(a) if the proposition is correct when $n = 1$; and
(b) if k is any value of n for which the proposition is correct, then the proposition is also correct for $n = k + 1$.

Now for the proof of some of the relationships that were accepted intuitively in the chapters.

1. (p. 33) Listing the remainders of the first few Fibonacci numbers upon dividing by 11, we have

1, 1, 2, 3, 5, 8, 2, 10, 1, 0, <u>1, 1, 2, 3, 5, 8, 2, 10, 1, 0,</u> ….

So we see that the remainders repeat in cycles of length 10. Since it is the remainder upon dividing a number by 11 that determines its divisibility by 11, all we have to do is check that in adding any ten consecutive numbers in the sequence

1, 1, 2, 3, 5, 8, 2, 10, 1, 0, 1, 1, 2, 3, 5, 8, 2, 10, 1, 0, ...

we get a sum divisible by 11. And we check this as follows. Since the cycle of this sequence is of length exactly 10, adding any ten consecutive numbers in this sequence will always come out to adding the ten numbers—1, 1, 2, 3, 5, 8, 2, 10, 1, 0—in a cycle.

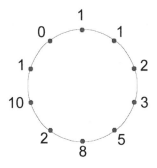

Figure B-2

(Imagine these ten numbers arranged clockwise in that order around a circle, with the sequence above obtained by traveling around the circle over and over. Then you can see that any numbers missed at the beginning of a cycle—if the sum is started somewhere in the interior of the cycle—are regained from the next cycle; for example, consider the sum $5 + 8 + 2 + 10 + 1 + 0 + \underline{1 + 1 + 2 + 3}$. This is because no matter where you start counting on the circle, counting ten numbers clockwise around the circle will amount to counting all ten numbers, because that is exactly how many numbers there are.)

These ten numbers have a sum of 33, which is indeed divisible by 11.

2. (p. 33) We proceed by proof by contradiction. $F_1 = 1$ and $F_2 = 1$ are, of course, relatively prime. Now, assume F_k and F_{k+1} are relatively prime (our hypothesis). If F_{k+1} and F_{k+2}

had a common factor other than 1, say the factor b, then since $F_k = F_{k+2} - F_{k+1}$, F_k would have the factor b too. But then F_k and F_{k+1} would both have the factor b, and that would contradict our assumption that F_k and F_{k+1} are relatively prime (do not have any common factor other than 1). So F_{k+1} and F_{k+2} *cannot* have any common factors other than 1, which means they are relatively prime as well. This completes our proof by contradiction.

3. (p. 35) We have to prove that the Fibonacci numbers in composite-number positions—with the exception of the fourth Fibonacci number, $F_4 = 3$—are composite numbers.

We will show that, in general, if m is divisible by n, then F_m is divisible by F_n, or, in other words, that F_{nk} is divisible by F_n for any n and k.

We proceed by induction on k. For $k = 1$, we need to check that $F_{n \cdot 1}$ is divisible by F_n; and this is, of course, true. Now assume F_{np} is divisible by F_n (this is our induction hypothesis). We want to prove the statement for $k = p + 1$, that is, that $F_{n(p+1)}$ is divisible by F_n. Using the lemma[1] in number 8 (following later)

$$(F_{m+n} = F_{m-1}F_n + F_m F_{n+1}),$$

we get:

$$F_{n(p+1)} = F_{np+n} = F_{np-1}F_n + F_{np}F_{n+1}$$

which is divisible by F_n because both $F_{np-1}F_n$ and $F_{np}F_{n+1}$ are divisible by F_n (by the induction hypothesis, F_{np} is divisible by F_n, and so is $F_{np}F_{n+1}$). This completes the induction.

---

1. A subsidiary proposition that is used to prove another proposition; a helping theorem.

So, based on all this, if n is a composite number, that is, if we can write $n = ab$, where a and b are both different from 1, then we know that F_n is divisible by F_a (and also by F_b, for that matter). Now, for n greater than 2, F_n is greater than any Fibonacci number that comes before it (the Fibonacci sequence is increasing), so since $n = 4$ (which is greater than 2) is the first composite number, certainly F_a is different from F_n. Now we also need to make sure F_a is different from 1 to show that F_n is composite (a composite number must have some divisor other than 1 and itself). Now, $F_a = 1$ can only happen if $a = 1$ or $a = 2$. But a (and b) was assumed not to be equal to 1 (because n is composite), so the only possibility is $a = 2$. The only problem, then, can occur when $a = 2$ and $b = 2$, that is, when $n = 4$. Indeed, for $n = 4$, $F_n = 3$, which is not composite.

5. (p. 37) We proceed by induction. For $n = 1$, we need to check that $F_2 = F_3 - 1$; and this is, indeed, true ($1 = 2 - 1$). Now assume the statement is true for $n = k$, that is, assume:

$$F_2 + \ldots + F_{2k} = F_{2k+1} - 1$$

Then we need to check the case when $n = k + 1$:

$$
\begin{aligned}
(F_2 + \ldots &+ F_{2k}) + F_{2k+2} \\
&= F_{2k+1} - 1 + F_{2k+2} \\
&= (F_{2k+1} + F_{2k+2}) - 1 \\
&= F_{2k+3} - 1 \\
&= F_{2(k+1)+1} - 1
\end{aligned}
$$

which is exactly the statement for $n = k + 1$. So our proof by induction is complete.

6. (p. 38) We again we proceed by induction. For $n = 1$, we need to check that $F_1 = F_2$; and this is, of course, true

($1 = 1$). Now assume the statement is true for $n = k$, that is, assume:

$$F_1 + \ldots + F_{2k-1} = F_{2k}$$

Then we must check to see if this holds true for $n = k + 1$:

$$\left(F_1 + \ldots + F_{2k-1}\right) + F_{2k+1}$$
$$= F_{2k} + F_{2k+1}$$
$$= F_{2k+2}$$
$$= F_{2(k+1)}$$

which is exactly the statement for $n = k + 1$. So our proof by induction is complete.

7. (p. 39) We again proceed by induction. For $n = 1$, we need to check that $F_1^2 = F_1 F_2$; and this is, of course, true ($1^2 = 1 \cdot 1$). Now assume the statement is true for $n = k$, that is, assume:

$$F_1^2 + \ldots + F_k^2 = F_k F_{k+1}$$

Then we check the case for $n = k + 1$:

$$\left(F_1^2 + \ldots + F_k^2\right) + F_{k+1}^2$$
$$= F_k F_{k+1} + F_{k+1}^2$$
$$= F_{k+1}\left(F_k + F_{k+1}\right)$$
$$= F_{k+1} F_{k+2}$$

which is exactly the statement for $n = k + 1$. So our proof by induction is complete.

8. (p. 42) We have, by factoring (difference of squares),

$$F_k^2 - F_{k-2}^2$$
$$= (F_k - F_{k-2})(F_k + F_{k-2})$$
$$= F_{k-1}(F_k + F_{k-2})$$
$$= F_{k-1}F_k + F_{k-1}F_{k-2}$$
$$= F_{k-1}F_{k-2} + F_k F_{k-1}$$

First we'll prove the following lemma, which will come in useful and later on.

LEMMA. $F_{m+n} = F_{m-1}F_n + F_m F_{n+1}$

Proof of Lemma

We proceed by induction on n. (Actually, this will be a form of induction that is called "strong induction," where we will be using the statements for $n = k - 1$ and for $n = k$ to prove the statement for $n = k + 1$. This will also mean that we will need two base cases rather than just one; so we will have to check the statement for $n = 1$ and for $n = 2$.) For $n = 1$, we have to check that $F_{m+1} = F_{m-1}F_1 + F_m F_2$, or, since $F_2 = 1$ and $F_2 = 1$, we have to check that $F_{m+1} = F_{m-1} + F_m$; this is, of course, true because it is the very relation by which we define the Fibonacci numbers. For $n = 2$, we have to check that $F_{m+2} = F_{m-1}F_2 + F_m F_3$, or, since $F_2 = 1$ and $F_3 = 2$, we have to check that $F_{m+2} = F_{m-1} + 2F_m$; and this is true by the following sequence of equalities:

$$F_{m-1} + 2F_m = (F_{m-1} + F_m) + F_m = F_{m+1} + F_m = F_{m+2}$$

Now assume the statement is true for $n = k - 1$ and $n = k$, that is, assume

$$F_{m+k-1} = F_{m-1}F_{k-1} + F_m F_k$$

and

$$F_{m+k} = F_{m-1}F_k + F_m F_{k+1}$$

(This is our induction hypothesis.) Then

$$F_{m-1}F_{k+1} + F_m F_{k+2}$$
$$= F_{m-1}(F_{k-1} + F_k) + F_m(F_k + F_{k+1})$$
$$= F_{m-1}F_{k-1} + F_{m-1}F_k + F_m F_k + F_m F_{k+1}$$
$$= F_{m-1}F_k + F_m F_{k+1} + F_{m-1}F_{k-1} + F_m F_k$$
$$= F_{m+k} + F_{m+k-1}$$
$$= F_{m+k+1}$$

That is, $F_{m-1}F_{k+1} + F_m F_{k+2} = F_{m+k+1}$, which is exactly the statement for $n = k + 1$. So our induction is complete.

Now using this lemma, we see (by substituting $m = k$ and $n = k - 2$) that

$$F_{k-1}F_{k-2} + F_k F_{k-1} = F_{k+k-2} = F_{2k-2}$$

Hence $F_k^2 - F_{k-2}^2 = F_{2k-2}$, and the proof is complete.

9. (p. 43) $F_n^2 + F_{n+1}^2 = F_{2n+1}$; that is, the sum of the squares of the Fibonacci numbers in positions n and $n + 1$ (consecutive positions) is the Fibonacci number in place $2n + 1$.

Proof

Using the lemma from item 8 (above), we prove our original statement. In the lemma, let $m = n + 1$ and $n = n$. Then we get $F_{2n+1} = F_n F_n + F_{n+1}F_{n+1}$, or, in other words, $F_{2n+1} = F_n^2 + F_{n+1}^2$, which is the equation we wanted.

10. (p. 43) A proof that $F_{n+1}^2 - F_n^2 = F_{n-1} \cdot F_{n+2}$

It follows immediately, because of the formula $[a^2 - b^2 = (a+b)(a-b)]$ and the definition of the Fibonacci numbers, that

$$F_{n+1}^2 - F_n^2 = (F_{n+1} + F_n)(F_{n+1} - F_n) = F_{n+2} \cdot F_{n-1}$$

11. (p. 44) A proof of $F_{n-1}F_{n+1} = F_n^2 + (-1)^n$, where n ≥ 1.
    By mathematical induction:

For $n = 1$:

$$P(1): \quad F_0 F_2 = F_1^2 + (-1)^1;$$
$$0 \cdot 1 = 1^2 - 1 = 0$$

For $n = k$:

$$P(k): \quad F_{k-1}F_{k+1} = F_k^2 + (-1)^k, \text{ where } k \geq 1.$$

For $n = k + 1$:

$$P(k+1): \text{ to check if it is true for } k+1:$$

$$
\begin{aligned}
F_{k+2}^2 - F_{k+1}^2 &= \left(F_k + F_{k+1}\right)^2 - F_{k+1}^2 \\
&= F_k^2 + 2F_{k+1}F_k + F_{k+1}^2 - F_{k+1}^2 \\
&= F_k \left(F_k + 2F_{k+1}\right) \\
&= F_k \left[\left(F_k + F_{k+1}\right) + F_{k+1}\right] \\
&= F_k \left(F_{k+2} + F_{k+1}\right) \\
&= F_k \cdot F_{k+3}
\end{aligned}
$$

And we will complete the proof by mathematical induction if we in the following use the assumption for $n = k$ in the form $F_{k-1} \cdot F_{k+1} - F_k^2 = (-1)^k$.

$P(k+1)$: To check if $P(k+1)$ is true : $F_k F_{k+2} = F_{k+1}^2 + \left(-1\right)^{k+1}$

$$F_k F_{k+2} - F_{k+1}^2 = F_k(F_{k+1} + F_k) - F_{k+1}^2$$
$$= F_k F_{k+1} + F_k^2 - F_{k+1}^2$$
$$= F_k^2 + F_{k+1}(F_k - F_{k+1})$$
$$= F_k^2 + F_{k+1}(-F_{k-1})$$
$$= F_k^2 - F_{k+1} F_{k-1}$$
$$= (-1)(F_{k+1} F_{k-1} - F_k^2)$$
$$= (-1)(-1)^k$$
$$= (-1)^{k+1}$$

Therefore, $P(k+1)$: $\ F_k F_{k+2} - F_{k+1}^2 = \left(-1\right)^{k+1}$

or $F_k F_{k+2} = F_{k+1}^2 + \left(-1\right)^{k+1}$

12. (p. 47) We have to show that F_{mn} is divisible by F_m for any n and m. We proceed by induction on n.

   For $n = 1$, we need to check that $F_{1 \cdot m}$ is divisible by F_m; and this is, of course, true. Now assume F_{mp} is divisible by F_m (this is our induction hypothesis—the statement for $m = p$). We want to prove the statement for $m = p + 1$, that is, that $F_{m(p+1)}$ is divisible by F_m. Using the Lemma in number 8, we get:

$$F_{m(p+1)} = F_{mp+m} = F_{mp-1} F_m + F_{mp} F_{m+1}$$

which is divisible by F_m because both summands (addends) $F_{mp-1} F_m$ and $F_{mp} F_{m+1}$ are divisible by F_m (by the induction hypothesis, F_{mp} is divisible by F_m), so it is also $F_{m(p+1)}$. This completes the induction.

14. (p. 49) We again proceed by induction. For $n = 1$, we need to check that $L_1 = L_3 - 3$; and this is, of course, true

$(1 = 4 - 3)$. Now assume the statement is true for $n = k$, that is, assume:

$$L_1 + L_2 + \ldots + L_k = L_{k+2} - 3$$

Then we must check to see if this holds true for $n = k + 1$:

$$L_1 + L_2 + \ldots + L_k + L_{k+1} = L_{k+3} - 3$$

We have

$$L_1 + L_2 + \ldots + L_k + L_{k+1}$$
$$= L_{k+2} - 3 + L_{k+1}$$
$$= L_{k+1} + L_{k+2} - 3$$
$$= L_{k+3} - 3$$

which is exactly the statement for $n = k + 1$. This completes our proof by induction.

15. (p. 51) We again proceed by induction. For $n = 1$, we need to check that $L_1^2 = L_1 L_2 - 2$; and this is of course true ($1^2 = 1 \cdot 3 - 2 = 1$). Now assume the statement is true for $n = k$, that is, assume:

$$L_1^2 + L_2^2 + \ldots + L_k^2 = L_k L_{k+1} - 2$$

Then we must check to see if this holds true for $n = k+1$:

$$L_1^2 + L_2^2 + \ldots + L_k^2 + L_{k+1}^2 = L_{k+1} L_{k+2} - 2$$
$$L_1^2 + L_2^2 + \ldots + L_k^2 + L_{k+1}^2$$
$$= L_k L_{k+1} - 2 + L_{k+1}^2$$
$$= L_{k+1} \left(L_k + L_{k+1} \right) - 2$$
$$= L_{k+1} L_{k+2} - 2$$

which is exactly the statement for $n = k + 1$. This completes our proof by induction.

For Chapter 4

Heron of Alexandria's Method for Constructing the Golden Section (figure 4-7) (p. 121)

$$AB = r, \ BC = CD = \frac{r}{2}, \text{ and } AD = AE = x$$

Applying the Pythagorean theorem to $\triangle ABC$:

$$AC^2 = AB^2 + BC^2$$

Therefore $\left(AD + CD\right)^2 = \left(x + \frac{r}{2}\right)^2 = r^2 + \left(\frac{r}{2}\right)^2$, or

$$x^2 + rx + \frac{r^2}{4} = r^2 + \frac{r^2}{4}, \text{ or } x^2 + rx - r^2 = 0$$

$$x_{1,2} = -\frac{r}{2} \pm \sqrt{\frac{r^2}{4} + r^2} = -\frac{r}{2} \pm \sqrt{\frac{5r^2}{4}} = -\frac{r}{2} \pm \frac{r}{2}\sqrt{5} = -\frac{r}{2}(1 \pm \sqrt{5})$$

Since x must be positive, we disregard $x = -\frac{r}{2}(1 + \sqrt{5})$; there-

fore $x = -\frac{r}{2}(1 - \sqrt{5}) = \frac{\sqrt{5} - 1}{2} \cdot \frac{r}{2} = \frac{1}{\phi} \cdot \frac{r}{2}$, or $x \approx .618033988 \cdot \frac{r}{2}$.

$$\frac{AE}{BE} = \frac{x}{r - x} = \frac{\dfrac{\sqrt{5}-1}{2} \cdot \dfrac{r}{2}}{r - \dfrac{\sqrt{5}-1}{2} \cdot \dfrac{r}{2}} = \frac{\dfrac{\sqrt{5}-1}{2} \cdot \dfrac{r}{2}}{\dfrac{4r}{4} - \dfrac{\sqrt{5}-1}{2} \cdot \dfrac{r}{2}} = \frac{(\sqrt{5}-1) \cdot \dfrac{r}{4}}{(4 - \sqrt{5} + 1) \cdot \dfrac{r}{4}}$$

$$= \frac{\sqrt{5}-1}{3-\sqrt{5}} = \frac{\sqrt{5}-1}{3-\sqrt{5}} \cdot \frac{3+\sqrt{5}}{3+\sqrt{5}} = \frac{3\sqrt{5}+5-3-\sqrt{5}}{9-5} = \frac{2\sqrt{5}+2}{4} = \frac{2(\sqrt{5}+1)}{4} = \frac{\sqrt{5}+1}{2}$$

$$= \phi \approx 1.618033988$$

$$AB = r, \ AE = BE = BC = CD = \frac{r}{2}, \text{ and } AD = AE = x = \frac{\sqrt{5}-1}{2} \cdot \frac{r}{2}$$

Construction of the Golden Section as in Figure 4-8 (p. 122)

$$AB = a, \ AC = \frac{a}{2}, \ BC = CD, \text{ and } AD = AE = x$$

Applying the Pythagorean theorem to $\triangle ABC$: $BC^2 = AB^2 + AC^2$

Therefore,

$$BC = CD = \sqrt{\frac{5a^2}{4}} = a\frac{\sqrt{5}}{2},$$

$$x = AE = AD = CD - AC = a\frac{\sqrt{5}}{2} - \frac{a}{2} = a \cdot \frac{\sqrt{5}-1}{2}, \text{ and}$$

$$BE = a - x = a\frac{3-\sqrt{5}}{2}$$

$$AE : BE = x : (a-x) = \frac{a \cdot \dfrac{\sqrt{5}-1}{2}}{a \cdot \dfrac{3-\sqrt{5}}{2}} = \frac{\sqrt{5}-1}{3-\sqrt{5}}$$

$$= \frac{\sqrt{5}-1}{3-\sqrt{5}} \cdot \frac{3+\sqrt{5}}{3+\sqrt{5}} = \frac{3\sqrt{5}+5-3-\sqrt{5}}{9-5} = \frac{2\sqrt{5}+2}{4} = \frac{2(\sqrt{5}+1)}{4} = \frac{\sqrt{5}+1}{2}$$

$$= \phi \approx 1.618033988$$

$$(AB = a, \ AC = \frac{a}{2}, \ BC = CD, \text{ and } AD = AE = x = a \cdot \frac{\sqrt{5}-1}{2})$$

Construction of the Golden Section as in Figure 4-9 (p. 123)

Begin by constructing a right $\triangle ABC$ with $AB = 2$ cm, $BC = 1$ cm. By the Pythagorean theorem, we get $AC = \sqrt{5}$ cm.

Using the theorem that states that the angle bisector of a triangle divides the side to which it is drawn proportionally to the other two sides, we find that

$$\frac{AP}{PB} = \frac{\sqrt{5}}{1}$$

and angle bisector $\overline{CQ}$ divides side $\overline{AB}$ (externally) proportionally to the sides of the angle to give us

$$\frac{AQ}{QB} = \frac{\sqrt{5}}{1}$$

Our task now is to show that point P divides the line segment $\overline{AB}$ into the golden ratio (or its reciprocal).

From the relationship established above, we get:

$$\frac{1}{\sqrt{5}} = \frac{PB}{AP} = \frac{PB}{AB - PB} = \frac{PB}{2 - PB}$$

$$PB\sqrt{5} = 2 - PB$$

$$PB = \frac{2}{\sqrt{5} + 1} = \frac{\sqrt{5} - 1}{2}$$

which is the reciprocal of the golden ratio ($PB = \frac{1}{\phi}$).

We can show that the point Q divides the line segment $\overline{AB}$ externally into the golden ratio.

From the second relationship above we get:

$$\frac{1}{\sqrt{5}} = \frac{QB}{AQ} = \frac{QB}{AB + QB} = \frac{QB}{2 + QB}$$

$$QB\sqrt{5} = 2 + QB$$

$$QB = \frac{2}{\sqrt{5} - 1} = \frac{\sqrt{5} + 1}{2}$$

which is the golden ratio (ϕ).

You might notice that in right ΔPCQ, CB is the mean proportional between PB and QB so that

$$\frac{CB}{PB} = \frac{QB}{CB}$$

$$\frac{1}{\dfrac{\sqrt{5}-1}{2}} = \frac{\dfrac{\sqrt{5}+1}{2}}{1}$$

which is certainly true!

For Chapter 6

Sums of Distinct Fibonacci Numbers (p. 188)

Every positive integer n can be expressed as a finite sum of distinct Fibonacci numbers.

Proof

Let F_k be the largest Fibonacci number $\le n$. Then $n = F_k + n_1$, where $n_1 \le F_k$. Let F_{k_1} be the largest Fibonacci number $\le n_1$. Then $n = F_k + F_{k_1} + n_2$, $n \ge F_k > F_{k_1}$.

Continuing like this, we get $n = F_k + F_{k_1} + F_{k_2} + \ldots$, where $n \ge F_k > F_{k_1} > F_{k_2} \ldots$. Since this sequence of decreasing positive integers must terminate, the result follows.

Proof of Fibonacci Generation of Pythagorean Triples (p. 192)

Suppose a, b, c, and d form a Fibonacci sequence. Then we get $c = a + b$, $d = c + b = a + b + b = a + 2b$. That is, a, b, $a + b$, and $a + 2b$ form a Fibonacci sequence.

The operations indicated in the procedure are:

$$A = \mathbf{2bc} = 2b(a + b) = 2ab + 2b^2$$
$$B = \mathbf{ad} = a(b + a + b) = a^2 + 2ab$$
$$C = \mathbf{b^2 + c^2} = b^2 + (a + b)^2 = a^2 + 2ab + 2b^2$$

Applying the Pythagorean theorem to see if these triples fit the theorem:

$$A^2 = (2ab + 2b^2)^2 = 4a^2b^2 + 8ab^3 + 4b^4$$
$$B^2 = (a^2 + 2ab)^2 = a^4 + 4a^3b + 4a^2b^2$$
$$C^2 = (a^2 + 2ab + 2b^2)^2 = a^4 + 4a^3b + 8a^2b^2 + 8ab^3 + 4b^4$$

$$A^2 \quad + \quad B^2 \quad = \quad C^2$$

$$[4a^2b^2 + 8ab^3 + 4b^4] + [a^4 + 4a^3b + 4a^2b^2] = [a^4 + 4a^3b + 8a^2b^2 + 8ab^3 + 4b^4]$$

More generally, to find a square which remains a square when increased or decreased by d, we have to find three squares that form an arithmetical progression with a common difference d.

The squares of the numbers are:

$$\left(a^2 - 2ab - b^2\right)^2 = a^4 - 4a^3b + 2a^2b^2 + 4ab^3 + b^4$$
$$\left(a^2 + b^2\right)^2 = a^4 + 2a^2b^2 + b^4$$
$$\left(a^2 + 2ab - b^2\right)^2 = a^4 + 4a^3b + 2a^2b^2 - 4ab^3 + b^4$$

They form an arithmetic progression with difference

$$d = 4a^3b - 4ab^3 = 4ab\left(a^2 - b^2\right)$$

If $a = 5$ and $b = 4$, then $d = 720$, and we have three squares $(a^2 + b^2)^2 = 41^2$, $41^2 - 720 = 31^2$, and $41^2 + 720 = 49^2$.

Dividing by 12^2, we have Fibonacci's solution.

From any one solution we can form infinitely many other solutions. Therefore, if we take $a = 41^2 = 1{,}681$ and $b = 720$, we have another solution, with

$$d = 5(24 \cdot 41 \cdot 49 \cdot 31)^2 = 11{,}170{,}580{,}662{,}080$$

and

$$(a^2 + b^2)^2 = 11{,}183{,}412{,}793{,}921$$

The three squares in arithmetic progression are then $x - 5$, x, and $x + 5$, with

$$x = \left(\frac{11{,}183{,}412{,}793{,}921}{2{,}234{,}116{,}132{,}416} \right)^2$$

Proof on the Property of 89 (pp. 216–17)

Refer to the development on pages 216–17.

$$10^{n+1} = 89 \cdot (F_1 \cdot 10^{n-1} + F_2 \cdot 10^{n-2} + \ldots + F_{n-1} \cdot 10 + F_n) + 10F_{n+1} + F_n \quad \text{(VI)}$$

We have already shown that (VI) is true for $n = 1$:

$$10^{1+1} = 89 \cdot (F_1 \cdot 10^{1-1}) + 10F_{1+1} + F_1 = 89 \cdot (1 \cdot 1) + 10 \cdot 1 + 1 = 89 + 10 + 1$$
$$= 100,$$

that is, $10^2 = 89 + 10 + 1$ \hfill (I)

Suppose that for some fixed integer $k \geq 1$,

$$10^{k+1} = 89 \cdot (F_1 \cdot 10^{k-1} + F_2 \cdot 10^{k-2} + \ldots + F_{k-1} \cdot 10 + F_k) + 10F_{k+1} + F_k$$

Then multiplying by 10 and again substituting (I), we get:

$$10^{k+2} = 89 \cdot (F_1 \cdot 10^k + F_2 \cdot 10^{k-1} + \ldots + F_k \cdot 10) + 10^2 F_{k+1} + 10 F_k$$

$$= 89 \cdot (F_1 \cdot 10^k + F_2 \cdot 10^{k-1} + \ldots + F_k \cdot 10) + (89 + 10 + 1) F_{k+1} + 10 F_k$$

$$= 89 \cdot (F_1 \cdot 10^k + F_2 \cdot 10^{k-1} + \ldots + F_k \cdot 10 + F_{k+1}) + 10 F_{k+1} + F_{k+1} + 10 F_k$$

$$= 89 \cdot (F_1 \cdot 10^k + F_2 \cdot 10^{k-1} + \ldots + F_k \cdot 10 + F_{k+1}) + 10 (F_{k+1} + F_k) + F_{k+1}$$

$$= 89 \cdot (F_1 \cdot 10^k + F_2 \cdot 10^{k-1} + \ldots + F_k \cdot 10 + F_{k+1}) + 10 F_{k+2} + F_{k+1},$$

since $F_{k+2} = F_{k+1} + F_k$.

This shows that (VI) is true in the $(k + 1)$st case, if it is true in the kst case.

And since we already have that it is true for $n = 1$, it is true for every positive integer n.

Dividing both sides of (VI) by $89 \cdot 10^{n+1}$, we obtain

$$\frac{1}{89} = \frac{F_1}{10^2} + \frac{F_2}{10^3} + \frac{F_3}{10^4} + \ldots + \frac{F_{n-1}}{10^n} + \frac{F_n}{10^{n+1}} + \frac{10 F_{n+1} + F_n}{10^{n+1}}$$

$$= \frac{F_1}{10^2} + \frac{F_2}{10^3} + \frac{F_3}{10^4} + \ldots + \frac{F_{n-1}}{10^n} + \frac{F_n}{10^{n+1}} + \frac{10 F_{n+1} + F_n}{10^{n+1}} + \ldots$$

where $\lim_{n \to \infty} \dfrac{10 F_{n+1} + F_n}{10^{n+1}} = 0$.

Chapter 9

Proof of the Binet Formula for Finding a Fibonacci Number (pp. 299–300)

The proof of the Binet formula $F_n = \dfrac{1}{\sqrt{5}} \left[\phi^n - \left(-\dfrac{1}{\phi} \right)^n \right]$ depends on finding expressons for ϕ^n and $\dfrac{1}{\phi^n}$. On page 114, in figure 4-2, we found expressions for ϕ^n for powers up to 10. These calculations suggested the formula

$$\phi^n = F_n\phi + F_{n-1} \tag{1}$$

We prove this by mathematical induction. For $n = 2$, the formula says $\phi^2 = F_2\phi + F_1 = \phi + 1$, which is true, as we saw on page 113.

Assuming (induction hypothesis) that the formula is true for $n = k$; that is,

$$\phi^k = F_k\phi + F_{k-1} \tag{2}$$

we show that it is true for $n = k + 1$:

$$\phi^{k+1} = F_{k+1}\phi + F_k$$

Multiply both sides of equation (2) by ϕ to get $\phi^{k+1} = F_k\phi^2 + F_{k-1}\phi$.

Since $\phi^2 = \phi + 1$, we have

$$\phi^{k+1} = F_k\phi^2 + F_{k-1}\phi = F_k(\phi+1) + F_{k-1}\phi = (F_k + F_{k-1})\phi + F_k = F_{k+1}\phi + F_k$$

as required.

In a similar way, we prove

$$\frac{1}{\phi^n} = (-1)^n\left(F_{n-1} - F_n \cdot \frac{1}{\phi}\right) \tag{3}$$

First, let's figure out $\dfrac{1}{\phi^2}$. We'll do this by using the formula $\phi^2 = \phi + 1$ twice.

$$\frac{1}{\phi^2} = \frac{1}{\phi+1} = \frac{1}{\phi+1} \cdot \frac{\phi-1}{\phi-1} = \frac{\phi-1}{\phi^2-1} = \frac{\phi-1}{(\phi+1)-1} = \frac{\phi-1}{\phi} = 1 - \frac{1}{\phi}$$

Now we verify that (3) holds for $n = 2$.

$$(-1)^2\left(F_1 - F_2 \cdot \frac{1}{\phi}\right) = 1 - \frac{1}{\phi} = \frac{1}{\phi^2}$$, which we've just found.

Assuming the induction hypothesis ($n = k$) that

$$\frac{1}{\phi^k} = (-1)^k\left(F_{k-1} - F_k \cdot \frac{1}{\phi}\right) \tag{4}$$

we show that

$$\frac{1}{\phi^{k+1}} = (-1)^{k+1}\left(F_k - F_{k+1}\cdot\frac{1}{\phi}\right)$$

Multiply both sides of equation (4) by $\frac{1}{\phi}$, to get

$$\frac{1}{\phi^{k+1}} = (-1)^k\left(F_{k-1}\cdot\frac{1}{\phi} - F_k\cdot\frac{1}{\phi^2}\right)$$

Since $\frac{1}{\phi^2} = 1 - \frac{1}{\phi}$, we have

$$\frac{1}{\phi^{k+1}} = (-1)^k\left(F_{k-1}\cdot\frac{1}{\phi} - F_k\cdot\left[1-\frac{1}{\phi}\right]\right)$$

$$= (-1)^k\left([F_{k-1} + F_k]\cdot\frac{1}{\phi} - F_k\right)$$

$$= (-1)^k\left(F_{k+1}\cdot\frac{1}{\phi} - F_k\right) = (-1)^{k+1}\left(F_k - F_{k+1}\cdot\frac{1}{\phi}\right)$$

completing the proof of equation (3).

To derive the Binet formula, we multiply both sides of equation (3) by $(-1)^n$, which enables us to rewrite it as

$$\frac{(-1)^n}{\phi^n} = \left(-\frac{1}{\phi}\right)^n = (-1)^{2n}\left(F_{n-1} - F_n\cdot\frac{1}{\phi}\right) = F_{n-1} - F_n\cdot\frac{1}{\phi}$$

(since -1 to an even power is $+1$).

Summarizing, we have equation (1) and the rewritten form of equation (3).

$$\phi^n = F_n\phi + F_{n-1}$$

$$\left(-\frac{1}{\phi}\right)^n = F_{n-1} - F_n\cdot\frac{1}{\phi}$$

If we subtract the second of these equations from the first, we obtain:

$$\phi^n - \left(-\frac{1}{\phi}\right)^n = F_n\phi + F_n\cdot\frac{1}{\phi} = F_n\left(\phi + \frac{1}{\phi}\right)$$

Since $\phi + \dfrac{1}{\phi} = \sqrt{5}$, the last equation can be written

$$\phi^n - \left(\dfrac{1}{\phi}\right)^n = \sqrt{5}F_n$$

Divide both sides by $\sqrt{5}$ and we're done.

Another Way to Find a Specific Fibonacci Number (Using a Calculator or a Computer) (p. 303)

We have to prove that $F_{2n} = F_n(2F_{n-1} + F_n)$.

We will use the following:

LEMMA. $F_{m+n} = F_{m-1}F_n + F_m F_{n+1}$

(This lemma was also used in the proof for chapter 1, item 8.)

We substitute $m = n$, then we get on the one side of the equation

$$F_{m+n} = F_{n+n} = F_{2n}$$

and on the other side

$$F_{m-1}F_n + F_m F_{n+1} = F_{n-1}F_n + F_n F_{n+1} = F_n\left(F_{n-1} + F_{n+1}\right)$$

also

$$F_{2n} = F_n\left(F_{n-1} + F_{n+1}\right)$$

Since $F_{n+1} = F_{n-1} + F_n$, we get:

$$F_{2n} = F_n\left(F_{n-1} + F_{n+1}\right) = F_n\left(F_{n-1} + F_{n-1} + F_n\right) = F_n\left(2F_{n-1} + F_n\right)$$

References

Alfred, Brother U. *An Introduction to Fibonacci Discovery*. San Jose, CA: Fibonacci Association, 1965.

Beutelspacher, Albrecht, and Bernhard Petri. *Der Goldene Sc hnitt*. Mannheim, BI-Wiss.-Verl., 1995.

Bicknell, Marjorie, and Verner E. Hoggath Jr., eds. *A Primer for the Fibonacci Numbers*. San Jose, CA: Fibonacci Association, 1973.

Blecke, Nathan. *Finding Fibonacci in Fractals*. MA thesis, Central Michigan University, 2001.

Boyer, Carl B. *A History of Mathematics*. New York: Wiley, 1991.

Burton, David M. *The History of Mathematics*. 3rd ed. New York: McGraw Hill, 1997.

Crownover, Richard M. *Introduction to Fractals and Chaos*. Boston, MA: Jones and Bartlett, 1995.

Dunlap, Richard A. *The Golden Ratio and Fibonacci Numbers*. River Edge, NJ: World Scientific Publishing, 1997.

Eatwell, John, Murray Milgate, and Peter Newman. *The New Palgrave*: *A Dictionary of Economics*. 4 vol. London: Basingstoke, 1987.

Eves, Howard. *An Introduction to the History of Mathematics*. Philadelphia: Saunders College Publishing, 1990.

Falconer, K. J. *The Geometry of Fractal Sets*. New York: Cambridge University Press, 1985.

Garland, Trudi Hammel. *Fascinating Fibonaccis*: *Mystery and Magic in Numbers*. Palo Alto, CA: Dale Seymour, 1987.

Ghyka, Matila. *The Geometry of Art and Life.* New York: Dover, 1977.

Heath, Sir Thomas L. *Manual of Greek Mathematics.* Oxford: Clarendon Press, 1931.

Herz-Fischler, Roger. *A Mathematical History of the Golden Number.* New York: Dover, 1998.

Hoggatt, Verner E., Jr. *Fibonacci and Lucas Numbers.* Boston: Houghton Mifflin, 1969.

Huntley, H. E. *The Divine Proportion.* New York: Dover, 1970.

Koshy, Thomas. *Fibonacci and Lucas Numbers with Applications.* New York: Wiley, 2001.

Mandelbrot, Benoît B. *The Fractal Geometry of Nature.* San Francisico: W. H. Freeman, 1982.

———. "A Multifractal Walk Down Wall Street." *Scientific American*, February 1999.

Posamentier, Alfred S. *Advanced Euclidean Geometry.* Emeryville, CA: Key College Publishing, 2002.

———. *Math Charmers: Tantalizing Tidbits for the Mind.* Amherst, NY: Prometheus Books, 2003.

Posamentier, Alfred S., and Ingmar Lehmann. *π: A Biography of the World's Most Mysterious Number.* Amherst, NY: Prometheus Books, 2004.

Prechter, Robert R., Jr. *R. N. Elliott's Masterworks: The Definitive Collection.* Gainesville, GA: New Classics Library, 1994.

Rasmussen, Steen Eiler. *Experiencing Architecture.* Cambridge, MA: MIT Press, 1964.

Rowley Kevin. "Fractals and Their Dimension." Plan B paper, Central Michigan University, 1996.

Runion, Garth E. *The Golden Section and Related Curiosa.* Glenview, IL: Scott Foresman, 1972.

Shishikura, M. "The Boundary of the Mandelbrot Set Has Hausdorff Dimension Two." *Astérisque* 222, no. 7 (1999): 389–405.

Sigler, Laurence. *The Book of Squares.* New York: Academic Press, 1987.

————. *Fibonacci's Liber Abaci*. New York; Springer-Verlag, 2002.

Vorobiev, Nicolai N. *Fibonacci Numbers*. New York: Blaisdell, 1961. Basel, Switzerland: Birkhäuser Verlag, 2002.

Walser, Hans. *The Golden Section*. Washington, DC: Mathematical Association of America, 2001.

Index

e. See Euler's number

φ. *See* golden ratio (section)

i. See Ludolph's number

0 (the number), 37, 50, 111, 205–206

2 (the number)
  Fibonacci numbers divisible by, 47, 210, 330
  power of, 196, 196n9

3 (the number)
  Fibonacci numbers divisible by, 47, 210, 330–31
  rule for divisibility by, 47n24

4 (the number)
  Fibonacci numbers divisible by, 330–31
  relationship to 89, 213n19

5 (the number), Fibonacci numbers divisible by, 47

6 (the number), Fibonacci numbers divisible by, 331

8 (the number), Fibonacci numbers divisible by, 48, 331

11 (the number), Fibonacci numbers divisible by, 33, 331, 350–51

12 (the number), Fibonacci numbers divisible by, 331

20 Steps around Globe (created by Niemeyer), 249–52, 250n21

89 (the number), 213–17, 213n19, 365–66

144 (the number), 208, 217, 328

666 (the number), 212–13

abacus, 12, 19n3

Acropolis (Athens), 232, 234

Adam and Eve (Dürer), 15, 268

Adam and Eve (Raimondi), 268

algebra, 12, 19–20, 19n2

algebraic number, 165n2

algorithms, 12, 198, 198n10, 310
  and ancient Egyptians, 195n8
  multiplication algorithm and Fibonacci numbers, 195–99

al-Khowarizmi, 19

alternating numbers, 42, 42n23, 44–45, 54, 54n25, 55, 142

Altevogt, Rudolf, 76

American Library Association, 211n16

angle
  angle sum of a triangle, 84
  bisectors, 123, 145, 146, 362
  divergence angle, 64n3
  and fractals, 311, 313, 314
  golden angle, 64, 64n3, 74, 148–49, 250, 250n20
  and isosceles triangles, 144
  and pentagons, 151, 157
  and right triangles, 128, 134, 135
  used in watch displays, 217–20, 220n21
  vertex angle, 144, 147

Animali da 1 a 55 (Merz), 252

antennas (in fractals), 322–25

Aphrodite of Melos (statue), 246–47

Apollo Belvedere (statue), 245–46

Arabic numerals. *See* Hindu numerals

architecture, Fibonacci numbers in, 232–44

arte povera, 252

arts and Fibonacci numbers, 231–69

Babbitt, Milton, 288

Babylonian formulae, 193

Bach, Johann Sebastian, 286

Baptism of Christ, The (Piero della Francesca), 268
Barbara, Saint, 260–61
baroque period, 277, 286
bar over digits, meaning of, 14, 14n3, 30, 108n1
Bartok, Bela, 285–88
base-sixty numbering system, 21, 21n6
base-ten numbering system, 11
base-two numbering system, 198
Bathers at Asnères (Seurat), 263
bear markets, 179, 182
bees, male, 13, 59–61
Beethoven, Ludwig van, 280–82
Béothy, Étienne, 247–48
Bernoulli, Daniel, 296
Bernoulli, Jacob, 131–32
Bernoulli, Nicolaus, I, 296
Bernoulli family, 296n3
"Bigollo." *See* Fibonacci, Leonardo
bijugate spirals, 69n8
binary forms of music, 274–75, 277
binary numbering system, 198
Binet, Jacques-Philippe-Marie, 28, 293–94, 296
Binet formula, 207
    for finding a particular Fibonacci number, 293–301, 328, 366–68
    for finding a particular Lucas number, 301–302
binomial coefficient, 88n3
binomial expansion, 87–88, 88n3
binomial theorem, 356
Birth of Venus, The (Botticelli), 259
bisectors of an angle, 123, 145, 146, 362
Bizet, George, 11
Bonacci, Guilielmo (William), 17–18
Bonacci, son of. *See* Fibonacci, Leonardo
Boncompagni, Baldassarre, 19n4
Bondone, Giotto di. *See* Giotto (di Bondone)
"Book of Squares." *See* Liber quadratorum (Fibonacci)
"book on calculation." *See* Liber Abaci (Fibonacci)
Borgliese, Pietro. *See* Piero della Francesca
Botticelli, Sandro, 259
Boulez, Pierre, 288
bracts, counting of, 13, 27, 63–64, 65–66
Braun, A., 27

Brouseau, Alfred, 65
Bruch, Hellmut, 254–55
Brunelleschi, Filippo, 239
bulbs (in fractals), 322–25
bull markets, 179
Bury, Claus, 255–56
business applications of Fibonacci numbers, 177–83

Cage, John, 290
calculator or computer used to find a Fibonacci number, 303–304, 368–69
Cantor, Georg, 308
cardiods (in fractals), 322–25, 322n7
Carmen (Bizet), 11
"Cartesian plane," 131
Cassai, Tommaso. *See* Masaccio
Cathedral of Chartres (France), 237–38
Center for International Light Art (Germany), 253–54
checkerboard, covering a, 188–91
Cheops (Khufu) Pyramid (Giza). *See* Great Pyramid (Giza)
Chinese remainder theorem, 24
Chopin, Frederic, 272–74
chords of a circle, 124, 124n14
chromatic scale, 286
Chu Shih-Chieh, 84n2
circles
    chords of, 124, 124n14
    concentric circles, 133
    congruent circles, 139–40
    and constructing a pentagon, 156–58
    great circle, 250, 250n19
    sequence of, 80–81, 84, 87
    used to construct golden ratio, 121–22, 124, 360–61
Circus Parade (Seurat), 263
Claus, M. *See* Lucas, François-Édouard-Anatole (Edouard)
coda in music, 281, 281n4
codetta in music, 281
coins and vending machines, 183–84
Collins, Charles, 178, 179
Cologne (Germany), 255–56
Columbia University, School of Library Economy, 211n16
common differences, 64, 78, 82, 330, 331, 364
common factors, 33–34, 33n20, 94, 192, 192n5, 351–52
See also relatively prime numbers

common fractions, 20n5, 24n12, 30

complex numbers, 120, 320

complex plane, 320, 320n5

composite numbers, 35, 35n21, 53–54, 352–53

Composition with Colored Areas and Gray Lines 1 (Mondrian), 269n36

Composition with Gray and Light Brown (Mondrian), 269n36

Composition with Red Yellow Blue (Mondrian), 269n36

compound interest, 24

computer or calculator used to find a Fibonacci number, 303–304, 368–70

computers and music, 289

concentric circles, 133

congruent circles, 139–40

consecutive numbers, 43–44, 55, 56

    four consecutive Fibonacci numbers, 211

    odd numbers, 295

    ratios of, 109–10, 210

construction

    constructing a pentagon, 155–58

    constructing fractals, 310–17

    constructing the golden ratio, 120–24, 362–65

continued fractions and Fibonacci numbers, 161–75, 162n1

    finite continued fractions, 163

    golden ratio as a continued fraction, 166–72

    infinite continued fractions, 163–64

Cook, Theodore Andrea, 245

corrective waves, 178, 179

Cossali, Pietro, 17n1

Couder, Yves, 74

credit cards, measurements of, 182

Crucifixion (Raphael), 268

crystallography, 342

cubits (as a measurement), 236

curves, 72, 131, 132–33, 247, 308, 322

    Gaussian curve, 246, 246n13

Curves of Life, The (Cook), 245

Dali, Salvador, 268

Davis, T. Antony, 76

decimal expansion, 108n2

De divina proportione (Pacioli), 245, 257, 260

denominator, 23

Deposition from the Cross (Weyden), 268

Der goldene Schnitt (sculpture by Ulrichs), 248–49

Der goldene Schnitt (Zeising), 115, 115n8

Der goldene Schnitt. Ein Harmoniegesetz und seine Anwendung (Hagenmaier), 248

Descartes, René, 131

Devaney, Robert, 319, 319n4

development in music, 276, 280, 281, 286

devil, sign of, 212, 213

Dewey, Melvil, 211n16, 271n1

Dewey Decimal classification system, 211, 211n16, 271n1

diagonal of the golden rectangle, 138–39

diatonic scales, 272

differences

    common differences, 64, 78, 82, 330, 331, 364

    Fibonacci differences, 78–80, 82, 83, 95n9

    in Lucas numbers, 301

    pattern in differences of squares, 43–44, 45–46, 54, 55, 357, 364

    sequences of progressive differences, 102

    sum and difference, 296–98

    and sum of successive powers, 297–98

digits

    curiosity of, 208

    first-digit patterns, 207–208

    last-digit patterns, 206–207

    See also integers; numbers

Di minor guisa (Fibonacci), 20

Dionysius' Procession (relief at Villa Albani), 267

distances, conversion of measures of, 200–203

divergence angle, 64n3

"divine proportion," 245

division

    common divisor, 337

    denominator, 23

    divisibility of Fibonacci numbers, 47–48, 55–56, 330, 341, 350, 358

    divisors of composite numbers, 353

    numerator, 23

    sequences of remainders, 31–32

dodecahedron, 231

Dodgson, Charles Lutwidge, 140

dominant in music, 285, 285n5
dominoes
    covering a checkerboard, 188–91
    knocking down as example of
        mathematical induction, 349–50
door, height of, 241
Douady, Stéphane, 74
"Dow theory" of investing, 178
drones, 13, 59–61
Duchamp, Gaston, 247, 247n14
Dudley, Underwood, 119
Dürer, Albrecht, 15, 155, 158, 258–59,
    259n26, 268
dyad in music, 285, 285n7
dynamic symmetry, 245, 245n10

Egyptian pyramids, 180, 234–37
Elements (Euclid), 20, 171
Elements of Dynamic Symmetry, The
    (Hambidge), 245
Elliott, Ralph Nelson, 177, 178–83
ellipse, area of, 133, 133n18
Empire State Building, climbing stairs
    of, 185
equal binary form of music, 274–75, 277
equations
    integer polynomial equation, 165n2
    linear equations, 23
    polynomial equation, 165n2
"equiangular spiral," 131
Essor II (sculpture), 248
Euclid, 12, 19, 20, 84, 171, 307
Euler, Leonhard, 120, 165n2, 296
Euler's number, 120, 165, 165n2
even-positioned Fibonacci numbers, 47,
    210, 330
    sum of, 37–38, 54, 353
event seating and Fibonacci numbers,
    221–22
Evolution: Progression and Symmetry
    III (Mields), 266–67
Evolution: Progression and Symmetry
    IV (Mields), 266–67
expansion, binomial, 87–88, 88n3
exposition in music, 280

factors, common, 33–34, 33n20, 192,
    192n5, 351–52
"factors of a multiplication," 23
"factors of a number," 23
Fatou, Pierre, 308
Fechner, Gustav, 115–17
feet (as a measurement), 241, 241n4

Feininger, Lyonel, 268
Fibonacci, Leonardo, 11, 17–22,
    23nn8–9, 56–57, 266, 327
Fibonacci Applications and Strategies
    for Traders (Fischer), 182
Fibonacci Association, 17, 28, 329
Fibonacci multiplication algorithm,
    198–99
Fibonacci Napoli (Merz), 252
Fibonacci Nim, 225–26
Fibonacci numbers, mathematical as-
    pects
    definition of, 53
    introduction to, 26–32
    list of first 500, 343–48
    and Lucas numbers, 15, 97–98,
        104–105, 227, 228–30, 297–98
    and the Pascal triangle, 90–98, 99,
        102, 103, 104–105
    proofs of Fibonacci relationships,
        349–58, 360–69
    properties of, 33–56, 337–41
    and Pythagorean triples, 192–94
    testing to determine if it is, 304–305,
        305n6
    See also golden angle; golden ratio
        (section); golden rectangle;
        golden spirals
Fibonacci numbers, use of
    in art and architecture, 231–69
    business applications of, 177–83
    climbing a staircase, 184–85
    converting miles and kilometers,
        200–203
    determining path of fish in a hatch-
        ery, 222–24
    determining seating at events,
        221–22
    and fractals, 307–25
    in geometry, 136–43
    in music, 271–91
    in nature, 12, 13, 25–27, 57, 59–76,
        132, 133
    in optics, 203–206
    and painting a house, 186
    and physics, 203–206
    and watch displays, 217–20
Fibonacci phyllotaxis, 64n3, 70–74
Fibonacci ratios, 94, 107–10, 180, 182
    See also golden ratio (section)
Fibonacci spirals. *See* golden spirals
Fibonacci's Temple (Bury), 255–56
Fifth Symphony (Beethoven), 280–82

finding a particular Fibonacci number in a fractal, 319, 319n4
    using a calculator or a computer, 303–304, 368–369
    using Binet formula, 293–301, 366–69
    using golden ratio, 303
finite continued fractions, 163
first differences. *See* differences
first digits of Fibonacci numbers, 207–208
Fischer, Robert, 182
fish, determining path in a hatchery, 222–24
fivefold symmetry element, 342
fixed integer, 365
Florence (Italy), 20, 238–40
Flos (Fibonacci), 20, 22
forbidden symmetry, 342
Four Books on Human Proportions (Dürer), 258
fourth differences. *See* differences
fractals, 307–25
fractions
    common fractions, 20n5, 24n12, 30
    continued fractions and Fibonacci numbers, 161–75, 162n1
    improper fractions, 161–62
    infinite continued fractions, 163–64
    proper fractions, 161, 162, 173
    unit fractions, 162
    use of horizontal bar in, 25
French Academy of Sciences, 200
French National Assembly, 200
Fridfinnsson, Hreinn, 266
fugues in music, 285–86, 287
furlongs (as a measurement), 200, 200n11

Galileo Galilei, 258
game of Fibonacci Nim, 225–26
Gauss, Carl Friedrich, 328
Gaussian curve, 246n13
Gelmeroda (Feininger), 268
generating fractals, 310–17
geodesic draft, 250n21
Geometric Compositions (Niemeyer), 249, 263–64
geometry, Fibonacci numbers in, 136–43
Gherardo, Giovanni di (da Prato), 239
Ghyka, Matil Costiescu, 243
Giotto (di Bondone), 17, 267, 267n29

Girl with the Ermine, The (da Vinci), 268
Giza, Great Pyramid at, 180, 234–37
golden angle, 148–49, 250, 250n20
    and plants, 64, 64n3, 74
golden ratio (section), 13–15, 13n2, 74–76, 107–60, 180, 263, 296, 327
    in art and architecture, 231–69
    and congruent circles, 139–40
    and consecutive Fibonacci numbers, 210
    constructing, 120–24, 360–63
    as a continued fraction, 166–72
    as an irrational number, 149–51
    and Lucas numbers, 174–75
    and measurements, 201
    in music, 271–91
    paper-folding and, 159
    powers of, 113–14
    precise value of, 111–12, 166n4
golden rectangle, 115–20, 123, 125–27, 128–29, 172
    in art and architecture, 232–33, 237, 259, 263
    and credit cards, 182
    diagonal of, 138–39
    and watch displays, 217–20
Golden Section, The (Hagenmaier), 248
golden spirals, 124–33, 263, 266
    See also spiral patterns
golden triangle, 107, 144–48, 260
    in pentagrams, 149–54, 158
great circle, 250, 250n19
Great Crash of 1929, 177
Great Pyramid (Giza), 180, 234–37
Grevsmühl, 250n21
Grimaldi, Giovanni Gabriello, 17n1
Gris, Juan, 269, 269n37
Grossman, George W., 313n3
Grossman truss, 313–19, 313n3

Hadrian's Arch, 234
Hagenmaier, Otto, 245, 248
Half a Giant Cup Suspended with an Inexplicable Appendage Five Meters Long (Dali), 268
Hambidge, Jay, 245
Hänsel und Gretel (Humperdinck), 11–12
harmonics in music, 283
Haydn, Franz Joseph, 278–80
Hermite, Charles, 165n2
Herodot, 235

Heron of Alexandria, 120–21, 362
hexagon, 155
Hindu numerals, 11, 11n1, 19, 23, 23n8
Hippasus of Metapontum, 149, 153
Hisâb al-jabr w'almuqabâlah (al-
	Khowarizmi), 19
Hispanus, Dominicus, 20
Hommage à Fibonacci (Bruch), 254–55
horizontal fraction bar, 25
human body and Fibonacci numbers, 74,
	75–76, 241–42, 257, 258
Humperdinck, Engelbert, 11–12
Hunter, J. A. H., 136

IBM, Watson Research Center, 320
Igloo Fibonacci (Merz), 253
immediate successor, 52
improper fractions, 161–62
impulsive waves, 178, 179
Indian figures. *See* Hindu numerals
induction and proofs of Fibonacci num-
	bers, 349–69
infinite continued fractions, 163–64
integer polynomial equation, 165n2
integers, 20n5, 21, 24n12, 33n20,
	112n7, 149, 196, 211
	fixed integers, 193
	and parity, 193, 193n6
	and pentagons, 152
	and perfect squares, 21n7
	positive integers, 48, 55, 162, 217,
		225, 331–32, 333, 336–37, 338,
		339, 363, 366
	and Pythagorean triangle, 193
	special integers, 114
	See also digits; numbers
investments and Fibonacci numbers,
	178–83
irrational numbers, 20, 20n5, 112,
	112n7, 149–51, 153, 163, 165n2,
	232, 300–301
isosceles triangles, 144–48, 151, 158,
	195, 261, 313, 314
iterations of fractals, 310, 311, 320–22,
	321n6
	in Grossman truss, 313–18
Ivy (Merz), 252

Japanese pagodas, 237
Jeanneret, Charles-Edouard. *See* Le
	Corbusier
Johannes of Palermo, 20
Johnson, Tom, 289–90

Journal of Recreational Mathematics,
	213
Julia, Gaston, 308, 320

Kepler, Johannes, 27, 119, 258
Khayyam, Omar, 20, 84n2
Khufu (Cheops) Pyramid (Giza). *See*
	Great Pyramid (Giza)
kilometers, conversion to miles,
	200–203
Knott, Ron, 91
Koch, Helge von, 310n1
Koch snowflake, 310–12

Lambert, Johann Heinrich, 165n2
Lamé, Gabriel, 29–30
Lamé numbers, 30
La Serie d'Or (Béothy), 247–48
Last digit of Fibonacci numbers,
	206–207
Last Supper, The (da Vinci), 268
leaf arrangements, 71–74
Leaning Tower of Pisa, 18, 23
Le Corbusier, 75, 76, 240–43, 244, 256
Leochares, 245–46, 246n12
Leonardo da Vinci, 74–75, 75n14, 257,
	260, 268
Leonardo of Pisa (Leonardo Pisano). *See*
	Fibonacci, Leonardo
Liber Abaci (Fibonacci), 11, 12, 19, 20,
	22–27
	See also Fibonacci numbers; rabbits,
		regeneration of
Liber quadratorum (Fibonacci), 20
linear equations, 23
logarithms
	and Euler's number, 120, 165n2
	logarithmic spiral, 69, 127, 131–32,
		133, 147, 182
Lonc, Frank A., 75–76
Lucas, François-Édouard-Anatole
	(Edouard), 15, 27–28, 27n15, 97
Lucas numbers, 15, 27–28
	Binet formula for finding a particular
		number, 301–302
	definition of, 53
	and Fibonacci numbers, 15, 97–98,
		104–105, 227, 228–30, 297–98
	and golden ratio, 174–75
	and golden spirals, 131
	Lucas spirals, 69n8
	and Pascal triangle, 97–104
	proofs for, 360–61

properties of, 49–53
as triangular numbers, 212
Ludolph's number, 120

Madachy, Joseph S., 171
Madonna Alba (Raphael), 261–62
Madonna and Child (Perugino), 268
Madonna Doni (Michelangelo), 268
main antenna (in fractals), 323
main cardioid (in fractals), 322–25
Mandelbrot, Benoît, 308–309, 320
Mandelbrot set, 319–25, 321n6
Masaccio, 267n30
mathematical induction and proofs of
    Fibonacci numbers, 349–69
measurements
    cubits, 236
    feet, 241, 241n4
    furlongs, 200, 200n11
    meters, 241, 241n4
    miles and kilometers, 200–203
Mersenne, Marin, 131
Mersenne prime numbers, 69, 69n7
Merz, Mario, 252–54
Mexican pyramids, 237
Michelangelo Buonarroti, 268, 268n34
Mields, Rune, 266–67
miles, conversion to kilometers,
    200–203
minor modulus, 331–36, 337–41
Modulon (Niemeyer), 264–65
*Modulor: A Harmonious Measure to the
    Human Scale Universally Applica-
    ble to Architecture and Mechanics,
    The* (Le Cobrusier), 241, 242, 256
Moivre, Abraham de, 296
Mona Lisa (da Vinci), 260
Mondrian, Piet, 269, 269n36
Mozart, Wolfgang Amadeus, 275–80,
    289
multijugate phyllotaxis, 70
multiplication, 23, 216, 225–26n24, 299,
    303, 312, 365–66
    finding a sum by multiplying, 41, 52,
        53
    and golden ratio, 114, 125
    and multiples of 6, 331
    multiplication algorithm and Fibo-
        nacci numbers, 195–99
    products of Fibonacci numbers,
        44–47, 56
    and Pythagorean triples, 192
    writing quotients, 30n19

music, Fibonacci numbers in, 271–91
*Music for Strings, Percussion, and Ce-
    lesta* (Bartok), 285–88

Narayan's Cows (Johnson), 290
natural numbers, 80n1, 120, 188n2,
    202n13, 207–208, 301n4, 350
    ordered sum of ones and twos, 187
    prime natural numbers, 193
    sequences of, 241
    as sum of Fibonacci numbers, 188
nature, Fibonacci numbers in, 12, 13,
    25–27, 57, 59–76, 132, 133,
    222–24
*Nature's Law—The Secret of the Uni-
    verse* (Elliott), 179, 180
nautilus shells, 132
"nest of radicals," 173
Neveux, Marguerite, 269
Niemeyer, Jo, 249–52, 263–65
nonconsecutive Fibonacci numbers, 188,
    202n13
nonprime numbers, 35, 35n21, 54
notation
    binary notation, 225
    and Fibonacci numbers, 37, 41,
        330
    and Lucas numbers, 50
    musical notation, 274
numbering systems
    base-10, 11
    base-60, 21, 21n6
    binary system (base-2), 198
    Hindu numerals, 11, 19, 23
    Roman numerals, 12
numbers
    algebraic number, 165n2
    alternating numbers, 42, 42n23,
        44–45, 54, 54n25, 55, 142
    complex numbers, 320, 320n5
    composite numbers, 35, 35n21,
        53–54, 352–53
    consecutive numbers, 43–44, 55, 56,
        109–10, 210, 211, 295
    irrational numbers, 20, 20n5, 112,
        112n7, 149–51, 153, 163,
        165n2, 232, 300–301
    natural numbers, 80n1, 120, 187–88,
        188n2, 193, 202n13, 207–208,
        241, 301n4, 350
    nonconsecutive Fibonacci numbers,
        187, 202n13
    nonprime numbers, 35, 35n21

odd prime numbers, 208, 328, 333–36, 337–41
palindromic numbers, 93–94, 93n6
pentatop numbers, 89
perfect numbers, 24, 24n11
prime natural numbers, 193
prime numbers, 33–34, 33n20, 35n22, 54, 69n7, 131, 211, 212, 331–37
primitive numbers, 339
rational numbers, 24, 24n12, 30, 163
reciprocals, 111, 113, 162, 169, 201, 214, 362
relatively prime numbers, 33–34, 33n20, 53, 193, 210, 228, 337, 338, 351–52
successive Fibonacci numbers, 209–11
symbol to show repetition of numbers, 14, 14n3, 30, 108n1
tetrahedral numbers, 89, 89n5, 90
transcendental numbers, 165n2
triangular numbers, 89, 89n4, 90, 211–13
See also digits; integers; squaring numbers
numerator, 23

odd-positioned Fibonacci numbers, sum of, 38–39, 54, 295, 355–56
odd prime numbers, 208, 328, 333–36, 337–41
optics and Fibonacci numbers, 203–206
ordered sum of ones and twos, 187

Pacioli, Luca, 243, 245, 257, 257n23, 260
pagodas, 237
painting a house and Fibonacci numbers, 186
Painting I (Mondrian), 269n36
paintings, Fibonacci numbers in, 257–69
palindromic numbers, 93–94, 93n6
Pankok, Otto, 269, 269n38
paper-folding and pentagon, 159
parastichies, 64–66
parity, 193, 193n6
Parthenon (Greece), 15, 111, 232–33, 245
Pascal, Blaise, 84, 84n2
"Pascaline" (calculator), 84
Pascal triangle, 84–105, 84n2

patterns
in differences of squares, 43–44, 45–46, 54, 55, 357, 366
of even Fibonacci numbers, 47
first-digit patterns, 207–208
last-digit patterns, 206–207
ordered sum of ones and twos, 187
of spirals in plants, 66–70
in squares of Fibonacci numbers, 42, 43–44, 45–46, 355–59
symbol to show repetition of numbers, 14, 14n3, 30, 108n1
"wave theory" of investing, 178–83
See also sequences
pentagon, 148, 148n23, 149–54, 231, 259, 259n26
constructing, 155–58
paper-folding and, 159
Pentagon (Washington, DC), 244
pentagram, 148, 149–54
paper-folding and, 159
pentatop numbers, 89
Pérez, José Victoriano González. *See* Gris, Juan
perfect numbers, 24, 24n11
perfect square, 21–22, 21n7, 78, 305, 305n6
Pericles, 232
"periodicity," 14, 14n3, 30, 108n1, 350
periodic decimals, 112n6
period of bulb or decoration (in fractals), 323–24
Persian War, 232
Perugino, Pietro, 268, 268n33
petal arrangements, 71, 74
Petrie, W. M. F., 236
Pfeiffer, Jörg, 250n21
Phidias, 110, 232
phyllotaxis, 64n3, 70–74, 70n10
physics and Fibonacci numbers, 203–206
Physicus, Theodorus, 20
π, 119–20, 165–66, 166n5, 237
Piano Sonatas (Haydn), 278–80
Piano Sonatas (Mozart), 277–78
Piero della Francesca, 268, 268n32
Pietro di Benedetto dei Franceschi (Pietro Borgliese). *See* Piero della Francesca
pineapples, 13, 63–64
pinecones, 13, 27, 64–66
Pisa (Italy), 18, 20, 22
Leaning Tower of Pisa, 18, 23

pitch generator, 290
plants, Fibonacci numbers in, 63–76
Plato, 153
Pollio, Vitruvius, 257–58, 257n24
Polo, Marco, 17
polynomial equation, 165n2
poor art, 252
positive integers, 55, 162, 217, 225, 331–32, 333, 336–37, 339, 363, 366
Practica geometriae (Fibonacci), 19
Prato (Giovanni di Gherardo da). *See* Gherardo, Giovanni di (da Prato)
Prelude No. 1 in C major (Chopin), 274
Prelude No. 9 in E major (Chopin), 273
Preludes (Chopin), 272–74
primary bulb (in fractals), 322, 324
prime numbers, 35n22, 54, 131, 211, 212, 331–37
    Mersenne prime numbers, 69, 69n7
    odd prime numbers, 208, 328, 333–36, 337–41
    prime natural numbers, 193
    relatively prime numbers, 33–34, 33n20, 53, 193, 210, 228, 337, 338, 351–52
    special primes, 336–37
primitive numbers, 339
products of multiplication, 44–47, 56
progressive differences, 102
proper fractions, 161, 162, 173
Propylaeum temple (Greece), 234
pyramids, 180, 234–37
Pythagoras and Pythagoreans, 149, 153, 231
Pythagorean theorem, 119, 119n11, 134, 235, 360
    using Fibonacci numbers to generate Pythagorean triples, 192–94, 363–65

quadratic formula, 113, 117, 117n10, 150
"queen of science," mathematics as, 328
Quetelet, Lambert Adolphe Jacques, 75
quotients, writing of, 30n19

rabbits, regeneration of, 12, 25–27, 57, 61
radicals, nest of, 173
Raimondi, Marcantonio, 268, 268n35
Raphael Santi, 260–62, 268
rational numbers, 24, 24n12, 30, 163

ratios
    Fibonacci ratios. *See* Fibonacci ratios
    golden ratio (section). *See* golden ratio (section)
    and irrational numbers, 149
    ratio of similitude, 128, 128n16
recapitulation in music, 276, 277, 280, 281
reciprocal rectangles, 128, 129
reciprocals, 111, 113, 169, 201, 214, 362
recreational mathematics, 24, 25, 27n15, 94, 94n7, 213
rectangles
    created using Fibonacci numbers, 40, 48–49
    created using Lucas numbers, 51
    golden rectangle, 115–20, 123, 125–27, 128–29, 138–39, 172, 182, 217–20, 232–33, 237, 259, 263
    puzzle concerning area of, 140–43
    reciprocal rectangles, 128, 129
    and triangles of equal area, 136–37
recursive sequence, 26, 27, 27n13, 31–32, 96–97, 185, 186, 293–94, 295, 296, 300
    in fractals, 310, 312
    and Lucas numbers, 97, 99
reflections in glass and Fibonacci numbers, 203–206
relatively prime numbers, 33–34, 33n20, 53, 193, 210, 228, 337, 338, 351–52
remainders, sequences of, 31–32
Rembrandt Harmenszoon van Rijn/nl, 268
Renaissance, 17, 231, 239, 245, 258, 267n30
ring, area of, 133
Roger de Le Pasture. *See* Weyden, Rogier van der
Roman mile (as a measurement), 200
Roman numerals, 12, 20
Romantic period, 280, 288
Rome, 237
Rubic's Cube, 94n7
rule for fractals, 320
rumors, distribution of, 61–62, 62n2
Russian peasant's method of multiplication, 195–99

Sacher, Paul, 285
St. Francis Preaching to the Birds (Giotto), 267

Salinger, J. D., 12
Santa Maria del Fiore Cathedral (Italy), 238–40
scales in music, 271–72, 282–83, 285nn5–6, 286
Schimper, C. F., 27
Schönberg, Arnold, 288–89
School of Athens, The (Raphael), 268
Scotus, Michael, 20
sculpture, Fibonacci numbers in, 244–56
seating at events and Fibonacci numbers, 221–22
second differences. *See* differences
Section d'Or (artist association), 247
seed of the fractal, 310–11, 320
Self-Portrait, A (Rembrandt), 268
sequences, 80n1
    of circles, 80–81, 84, 87
    and Fibonacci numbers, 77–84, 114, 174–75
    and Lucas numbers, 174–75
    of natural numbers, 241
    and Pascal triangle, 87–105
    of progressive differences, 102
    recursive sequence, 27, 27n13, 31–32, 96–97, 293–94
    of remainders, 31–32
    See also patterns
Sérusier, Paul, 269
Seurat, Georges, 262–63
sexagesimal number. *See* base-sixty numbering system
Sierpinski, Waclaw, 312n2
Sierpinski gasket, 312–13
Sigler, Laurence E., 19n4
Signac, Paul, 269
"Sign of the Devil, The" (Wang), 213
similitude, ratio of, 128, 128n16
Simson, Robert, 171
Singmaster, David, 94
Sistine Madonna (Raphael), 260–61
Sixtus II (pope), 260–61
snails, 133
sonata-allegro form of music, 275–76, 281
sonatas as a musical form, 275–80, 276n2
special integer, 114
special primes, 336–37
spiral patterns, 65–74, 69n8, 131
    artificial Fibonacci-flower spirals, 70
    bijugate spirals, 69n8
    "equiangular spiral," 131

golden spirals, 124–33, 263
logarithmic spiral, 127, 131–32, 133
and pentagons, 159–60
spokes (in fractals), 323
square roots, 305
squares
    created using Fibonacci numbers, 40, 48–49
    created using Lucas numbers, 51
squaring numbers, 293–96
    alternating numbers and, 44–47, 55, 142
    pattern in differences of squares, 43–44, 45–46, 54, 55, 355, 364
    patterns in squares of Fibonacci numbers, 42, 43–44, 45–46
    perfect square, 21–22, 21n7, 78, 305, 305n6
    subtracting squares of Fibonacci numbers, 42–43, 355–59
    sum of squares of Fibonacci numbers, 39–42, 43, 54, 353–54
    sum of squares of Lucas numbers, 51–53, 56
    transforming fractals, 320–21
    See also numbers
staircases, Fibonacci numbers and climbing, 184–85
static symmetry, 245
statute mile (as a measurement), 200
stock market and Fibonacci numbers, 178–83
Stonehenge (Great Britain), 237
Stradivarius, Antonio, 291
subtracting squares of Fibonacci numbers, 42–43, 357–61
successive Fibonacci numbers, 209–11
sum
    angle sum of a triangle, 84
    and difference of successive powers, 297–98
    of even-positioned Fibonacci numbers, 37–38, 54, 353
    of Fibonacci numbers and distance conversion, 202, 202n13
    Fibonacci numbers in Pascal triangle, 91–94
    finding a sum by multiplying, 41, 52, 53
    formulas for getting sum of Fibonacci numbers, 36–42, 53, 54
    formulas for getting sum of Lucas numbers, 49–51, 56

natural numbers as sum of Fibonacci
    numbers, 188, 365
of odd-positioned Fibonacci num-
    bers, 38–39, 54, 295, 353–54
ordered sum of ones and twos, 187
in Pascal triangle, 85–87
sequences of sums of Fibonacci
    numbers, 82–83
sum and difference, 296–98
sum of squares of Fibonacci num-
    bers, 39–42, 43, 54, 353–54
sum of squares of Lucas numbers,
    51–53, 56
summand, 225, 225n23
sunflowers, 68–69, 69n8
Swiss Guard, uniforms for, 268n34
symbol to show repetition of numbers,
    14, 14n3, 30, 108n1
symmetry, 245, 245n10, 266, 342
    forbidden symmetry, 342

Temple of Olympia, 111
testing numbers to find Fibonacci num-
    bers, 304–305, 305n6
test of numbers in fractals, 320, 321,
    321n6
tetrahedral numbers, 89, 89n5, 90
thermodynamics, laws of, 245n10
39th Mersenne prime number, 69, 69n7
Thorndike, Edward Lee, 117
tonic in music, 285, 285n6
Tower of Hanoi puzzle, 27n15
transcendental numbers, 165n2
triangles
    angle sum of a triangle, 84
    and constructing the Grossman truss,
        314–19
    Fibonacci numbers as lengths of
        sides, 195
    golden triangle, 144–48, 149–54
    isosceles triangles, 144–48, 314
    Omar Khayyam and, 84n2
    Pascal triangle, 84–105
    Pythagorean triangles and Pythago-
        rean triples, 192–94
triangular numbers, 89, 89n4, 90,
    211–13, 212n17
Trinity (Masaccio), 267, 267n30

triples, Pythagorean, 192–94, 365–67
Tristan und Isolde (Wagner), 282–85
Triumph of Galatea, The (Raphael), 268
twelve-tone method in music, 288–89
20 Steps around Globe (created by Nie-
    meyer), 249–52, 250n21

Ulrichs, Timm, 248–49
unequal binary form of music, 274–75
United Nations headquarters (New York
    City), 243
unités d'habitation (France), 240
unit fractions, 162
UNIVAC (computer), 289, 289n8
Unna (Germany), 253–54
Untitled (Fridfinnsson), 266
Utsjoki (Niemeyer), 263–64

"Valuable Mirror of the Four Elements,
    The" (Khayyam), 84n2
Vannucci, Pietro. See Perugino, Pietro
Variation VI (Niemeyer), 264
vending machines and Fibonacci num-
    bers, 183–84
Venus de Milo (statue). See Aphrodite
    of Melos (statue)
vertex angle, 144, 147
Villon, Jacques, 247, 247n14
violin as example of golden ratio, 271,
    291
"Vitruvian" man, 257
Vitruvius, 257–58

Wagner, Richard, 282–85, 287
Wang, Steve C., 213
watches, displaying of, 217–20
Watson Research Center, IBM, 320
Wave Principle, The (Elliott), 179
"wave theory" of investing, 178–93
Webern, Anton von, 289
Weyden, Rogier van der, 268, 268n31
worker bees, 59

Zeckendorf, Edouard, 188n4, 202n13
Zeckendorf theorem, 188n4
Zeising, Adolph, 75, 115, 115n8, 243
zephyr (zero), 11, 19, 23
Zeus (statue of), 111